Rain Harvesting in the Rainforest: The Ancient Maya Agricultural Landscape of Calakmul, Campeche, Mexico

Helga Geovannini Acuña

BAR International Series 1879
2008

Published in 2016 by
BAR Publishing, Oxford

BAR International Series 1879

Rain Harvesting in the Rainforest: The Ancient Maya Agricultural Landscape of Calakmul, Campeche, Mexico

ISBN 978 1 4073 0359 8

BAR Publishing is the trading name of British Archaeological Reports (Oxford) Ltd. British Archaeological Reports was first incorporated in 1974 to publish the BAR Series, International and British. In 1992 Hadrian Books Ltd became part of the BAR group. This volume was originally published by Archaeopress in conjunction with British Archaeological Reports (Oxford) Ltd / Hadrian Books Ltd, the Series principal publisher, in 2008. This present volume is published by BAR Publishing, 2016.

Printed in England

BAR
PUBLISHING

BAR titles are available from:

BAR Publishing
122 Banbury Rd, Oxford, OX2 7BP, UK
EMAIL info@barpublishing.com
PHONE +44 (0)1865 310431
FAX +44 (0)1865 316916
www.barpublishing.com

TABLE OF CONTENTS

LIST OF FIGURES

LIST OF TABLES

SUMMARY

Rain Harvesting in the Rainforest: The Ancient Maya Agricultural Landscape of Calakmul, Campeche, Mexico

The main subject discussed in this study is the way in which the ancient Maya of Calakmul, who thrived between 900 B.C. to A.D. 1000, managed their landscape in order to survive in the tropical rainforest. Their lithic technology, the hot, humid climate with a prolonged dry season, the lack of permanent surface sources of fresh water, and thin soils, considered insufficient for sustained agricultural production, are factors that were addressed successfully by the Maya in developing their complex civilization.

Landscape, archaeological, and edaphological analyses are performed, after which the study explores the areas most advantageous to permanent habitation, suitable agricultural zones, land potential of the region and the capability of the area for supporting population. In addition, a complex agricultural channel irrigation system is developed as a critical factor for managing productive terrain for agriculture in karstic depressions (*bajos*). Similarly, an impressive rain harvesting system is exposed as an option to optimize hydrological resources for canalizing excessive rain during the wet season and storing water during the dry period. Finally, a reconstruction of the agricultural landscape is proposed.

RESUMEN

Colectando Lluvia en la Selva: El Paisaje Agrícola Maya del Sitio Arqueológico de Calakmul, Campeche, México.

El tópico principal que se aborda en esta tesis es la manera en que los antiguos mayas de Calakmul, que florecieron de 900 A.C. a A.D. 1000, manejaron su paisaje para sobrevivir en la selva tropical. Su tecnología lítica, el clima húmedo y cálido con una estación seca prolongada, la falta de fuentes permanentes de agua y los suelos delgados son factores que no les impidieron desarrollar una compleja civilización.

En este trabajo, a través del análisis del paisaje, se explora el potencial del terreno de Calakmul, las zonas adecuadas para la agricultura, para la habitación y la capacidad de la región para soportar una población numerosa. Adicionalmente se expone un sistema de riego potencial a través de canales de irrigación que fue desarrollado para aprovechar el terreno apto para la agricultura en depresiones cársticas (bajos). De manera similar, se analiza un impresionante sistema de colección de lluvia como la opción para optimizar la captación de agua durante la época de secas, y su canalización y almacenamiento durante la época de lluvias. Finalmente se propone una reconstrucción del paisaje agrícola de Calakmul durante los periodos mesoamericanos Preclásico y Clásico.

ACKNOWLEDGEMENTS

The accomplishment of this study was possible thanks to the involvement of many people. First of all, I appreciate the trust of Professor Peter Mathews. Under his supervision I was able to manage this project with complete freedom, particularly after some long-lasting adversities and unplanned situations. In the same way, I offer my thanks to the archaeologist Ramón Carrasco, director of the Calakmul Archaeological Project. He kindly opened the doors of his project, providing the permits and resources for staying and working at Calakmul. Thanks also to the examiners of this study, Dr. Armando Anaya (Universidad Autónoma de Campeche), Dr. Marcello Canuto (Yale University) and Dr. Rodrigo Liendo (Universidad Nacional Autónoma de México), for their comments and constructive critique.

Other individuals helped a lot in diverse ways. Dr. Mario Aliphat lent me valuable soil equipment, e-mailed me remarkable remote sensing images and gave valuable advice for this research. Archaeologists Alma Eugenia Martínez and Kai Delvendahl contributed with the identification of the ceramics of the excavated pits. Archaeologists Marinés Colón, Omar Rodríguez-Campero and Agustín Anaya facilitated the development of this project in field during the 2003 and 2004 seasons. Workers José Antonio Guzmán and Anselmo León assisted me with their hard labour during the first season, as well as Isidro Hernández and Eusebio Paredes during the second. Alberto Soria supported this project by driving the heavy equipment from Mexico City to Calakmul during the 2003 season. He also lent me a hand with the extraction of pollen and soil cores. Dr. Gerald Islebe of Ecosur, Chetumal, did preliminary pollen analyses of soil samples in his laboratory. Although there was not enough pollen for identification, he showed enthusiasm for the project. Edaphologist Hector Estrada, from University of California Riverside examined the chapter related with soils and made valuable observations. In Australia, I appreciate the support of Professor Tim Murray, Wei Ming, Rudy Frank, Rose Allen and Stella Bromilow. I also want to express my gratitude to Eduardo Perez de Heredia, Claudia García Solís, Eileen Uribe-Querol and Hector Trujillo for their valuable support in different stages of the study. In addition, Joshua Englehardt revised some sections of the document with great care. Finally, thanks to my family, Alberto, Sebastián and Eva for their constant support. To them I dedicate this work.

Field reconnaissance and laboratory analyses were completed thanks to La Trobe University Faculty of Humanities and Social Sciences Research Grant for Higher Degree Research Students. During the development of the research, I held the International Postgraduate Student Scholarship and La Trobe University Postgraduate Student Scholarship.

CHAPTER 1. QUESTIONS, OBJECTIVES, CONTRIBUTIONS AND FRAMEWORK

"...when we talk about relations between man and environment, we should observe with precision the degree or relation that exists between human population and its environment. In some cases, the environment with which population relates is physical (the nature), meanwhile in other cases, it will be formed principally by social institutions (it is to say, society). With the possible exception of primitive bands (hunter-gatherer societies), human communities depend on social mediation as much or more than on physical environment."

Morán, 1993:27.

Research Questions

A major question that has intrigued Maya researchers for years has been that of how the ancient Maya survived in such a harsh environment with their stone tool technology and developed such a complex society (e.g. Fedick, 1995; 1996; Harrison and Turner, 1978; Roys, 1943). The hot, humid tropical climate, limited surface water (almost inexistent in some areas), and the fact that the soils of the tropics have been considered to be infertile and incapable of supporting sustained agricultural production (Sánchez and Logan, 1992) are some previously identified constraints that the ancient Maya addressed in order to develop such an impressive civilization.

Cultivating in the Maya area, particularly the lowland region, represents a difficult task. Even today, with complex technological options, agriculturalists have avoided some regions (García Gil, 2002). In addition to the limitations mentioned above, the difficulties of navigating the terrain in the wet season, the constant weed invasion of the cultivated areas, and disadvantageous detrimental fauna, such as insects and serpents, have been transcendental factors that have impeded the practice of agriculture in both historical and modern times, since so much effort must be invested to obtain even the smallest harvest.

So, how was it possible for the Maya to survive in places that even with modern technological advances are so difficult to occupy? This is the main question that is tackled in this study. Following this line, issues such as the landscape elements that permitted the establishment of a settlement in that area will be analyzed. Concretely, the best places for inhabitation, suitable areas for agriculture and rainforest, the land potential of the area, the capability of the region for supporting food supplies for a large number of inhabitants are topics to pursue in the context of the general question of how ancient inhabitants conceived their landscape.

Advances in Maya Subsistence Research

Regarding the fundamental question of Maya subsistence, different hypotheses have been proposed. A heated debate began early in the last century. From the 1940s until 1960s, the majority of Maya scholars believed that slash and burn agriculture was the only cultivation option (e.g. Morley, 1947:143). It was thought that since maize had been the basic food for sixteenth century Mesoamerican inhabitants, then it was the most important crop for the ancient ones.

This vision has been called into doubt by evidence from other alternative agricultural practices (Culbert, 1995; Harrison and Turner, 1978; Hernández Xolocotzi, 1959); for example, selective agro-silviculture (Gómez Pompa et al., 1987; McKillop, 1994; Puleston, 1978), the presence of family orchards (Wiseman, 1978), horticulture (Dahlin, 1985) and wetland agriculture (Culbert et al., 1990; Pope and Dahlin, 1989). Taking advantage also of the forest and aquatic resources with alternative agricultural practices would have satisfied the requirements of large populations in the Yucatan Peninsula.

This vision has prevailed until today. International multidisciplinary teams using the latest technological advances have explored this hypothesis in detail. Innovative techniques, equipment and software have permitted a more complete picture of Maya agriculture. For example, the Central Petén area has a long history of palaeo-ecological research (Deevey et al., 1979). As part of the Central Petén Historical Ecology Project, the lakes Yaxha, Sacnab, Macanche, Salpeten, Quexil, Petenxil, Peten Itza, Oquevix, Ija, Chilonche and Chimaj have been thoroughly investigated (Rice, 1996). Research concerned with the ancient environment has been done in the Maya Lowlands (e.g. Adams, 1990; Curtis et al., 1998; Dunning et al., 1999; 2002; Hansen et al., 2002; Islebe et al., 1996) and at present it is a recurrent theme in Mesoamerican archaeology. Sites such as El Pilar, Río Azul, El Mirador, Nakbe and Tikal have been also extensively studied.

Although so much time and energy have been dedicated to the study of Maya subsistence, there are still many questions that have not been satisfactorily answered. For example, the elevated population density (about 200 per km^2) estimated in Maya Lowlands (Turner, 1990) would have depleted their environment with swidden agriculture in a short time span (Lundell, 1937). In the same way, large populations would have needed abundant water for human consumption and agriculture. This situation would have exhausted the natural resources in a short period.

In Maya area, there has been a tendency to universalize the functionality of each agricultural technique or subsistence option. For this reason, the proposed agricultural methods and techniques of ancient people should be studied according to the physical conditions of each site or at least a group of sites that share environmental and relief features (Dunning, 1996). For example, in the Yalahau Region, in northern Yucatan Peninsula, the Maya developed wetlands and managed the terrain using soil or algae to enrich upland garden plots. They also cultivated trees within their communities (Fedick and Morrison, 2004).

A subject intimately linked to the discussion of Maya subsistence is the so-called Maya Collapse. As one reviews agriculture literature, one inevitably encounters the possible causes for the disappearance of the Classic Maya civilization. In the ninth and tenth centuries A.D., structures and stelae ceased to be erected and the majority of the lowland cities were abandoned. After many decades of debate a uniform hypothesis has still not been thoroughly accepted. The Maya Collapse has been one of the most recurrent subjects, which accords Maya civilization an air of mystery. Impressively, Gill reports eighty-eight theories and their variants to explain the Classic Maya downfall (Gill, 2001). After reviewing palaeo-climatic and ethno-historical records, he concludes that the Collapse was caused by severe droughts. Only the Maya who found a way to stay close to water sources survived.

In sum, the Collapse has been explained by dramatic climatic modifications that promoted drought, which in turn, threatened the Maya way of life. As a consequence, deforestation, erosion, epidemics, lack of resources, and finally, starvation and death ended in the disintegration of the Classic Maya culture. (e.g. Binford et al., 1987; Curtis et al., 1996; Deevey et al., 1979; Gill 2001; Leyden et al., 1996; Rosenmeier et al., 2002).

The Archaeological Site of Calakmul

Located in the confluence between a karstic spine and a huge *bajo* (*polje* or large closed karstic depression) are the ruins of the ancient kingdom of Calakmul. This site is found in the south of the Mexican state of Campeche, about 50 kilometers west of the border of the state of Quintana Roo and about 30 kilometers north of the Department of Peten in Guatemala (See Figure 1.1). Calakmul thrived for two millennia from the Middle Preclassic to Terminal Classic (from ca. 900 B.C. to A.D. 1000). The people of Calakmul developed an impressive culture. The imposing pyramidal buildings, tombs, murals, palace complexes and infrastructure that we marvel at today are just a pale reflection of the magnificence of the Calakmul kingdom at its zenith.

Although this site was first documented in the 1930s (Lundell, 1933), major archaeological work at Calakmul began in the 1980s (e.g. Folan, 1985; Folan et al., 2003; Carrasco et al., 1999a; 2000; 2001; 2002), and had been virtually continuous since then. The excavation and restoration of multiple monumental buildings, the analyses of remains and the study of the environment have been invaluable elements in understanding Calakmul. Literally tons of ceramics and lithics have been recovered as a result of continuous excavations. Even though archaeologists have been actively communicating their results, there are still many interesting details that we do not know about this site.

Previous research related to the environment of the Biosphere of Calakmul (Díaz Gallegos, et al., 2002; Folan et al., 1999; 2001a; García Gil et al., 2002; Gunn et al., 2002) has been useful in obtaining a general vision of the ancient Calakmul landscape. However, I believe that a study which integrates all this information and considers the material evidence and proposes characteristics of the site landscape and its potential use could shed light on the subsistence options of its inhabitants. For this reason, this study was undertaken.

The Importance of Landscape

Day after day, humans modify the places they inhabit. From hunter-gatherer groups to crowded cities, people have continually shaped and re-shaped the earth's physical environment. The expression of this change and its rate has typified individuals or groups in a particular way. Human subsistence activities, social organization and religious thoughts have decisively molded their physical habitat.

But the human aspect is not the only factor that transforms environments. The landscape is the material or imaginary socio-cultural product that emerges from the influence of natural forces, living organisms and human manipulation on natural and created elements in specific spatial and temporal contexts (Criado, 1999). Hence, a landscape is a complex entity that exposes a history of human cultural practices.

Cultural practices are not predetermined by a specific place and its environmental conditions. The physical setting forms the "building blocks" that are subsequently arranged in a particular way. The landscape is the summation of the physical conditions of a place, such as relief or vegetation, of human activities and thoughts, and of the way they are manifested in the surroundings. Each cultural group modifies the same place in a particular way, as their options are limited by their technology, as well as social and religious practices (Criado, 1997).

When archaeologists encounter landscapes, they confront intricate issues they hope to be able to unravel. One of them is the reconstruction of the prior natural conditions and their historical succession. Another issue is the elucidation of the human influence in a place, given that a landscape is the present result of the historical influences of different people or groups in different temporal contexts. Finally, it should be remembered that a landscape connotes symbolic appropriations of nature (Ingold, 1986), and each researcher is just an interpreter that perceives the landscape with a very particular point of view, probably much different from the ancient landscape settlers. All of these issues are difficult to comprehend but paramount for a lucid understanding of the way humans relate and are related to their environment.

Figure 1.1. Location of Calakmul and other ancient Maya sites.

Despite the aforementioned constraints, the investigation of landscape remains an interesting and useful starting point for approaching the study of cultures. When archaeologists investigate landscape, they not only try to reconstruct past physical characteristics; rather, their main objective is to elucidate the implicit cultural complexity. Landscape archaeology constitutes a valuable framework for approximating this complexity.

The Aims of this Work

- General Objective

The main purpose of this work is to explicate the way in which the ancient Maya of Calakmul managed their landscape. For that end, a particular vision of this ancient landscape will be put forward, as well as proposing the plausible interactions of the inhabitants with their environment. Themes such as subsistence, agriculture, land potential, suitable crops, water management and adaptation to the physical surroundings will be discussed in the context of subsistence.

- Specific Objectives

1. To gather topographical, geographical, geological, soil, hydrological and climatic information of Calakmul to compile a geographical database. Published cartographic charts of the area, as well as aerial photographs and general geographic information have been compiled. The topographic maps 1:50 000 (INEGI, 1987), 1:20 000 (SIGSA, 2004) and 1:6250 (May Hau et al., 1990) of the area were digitized in order to obtain base maps for the subsequent analysis. Other maps previously generated by

other researchers were considered too (García Gil, 2002; Lundell, 1933; Ruppert and Denison, 1943) for the observation of the site and its context. In addition, aerial photographs scale 1:75 000 (INEGI, 2001a), ortho-photographs (INEGI, 2001b), a multi-band Aster image of Campeche, as well as a Landsat, Earthsat from ESRI (2004) and Terra (Skyline Software Systems 2003) images of Calakmul were taken into account for image analyses.

2. To identify the landscape units of the study area in order to establish the major differences through the physiognomic-lithomorphic method (Breimer et al., 1986). A landscape unit is the spatial combination of the topography, parent material and types of soils to form a minimal component. These units were identified as the basic units for landscape analysis. They were used for characterizing the landscape and for emphasizing its physical differences.

3. To identify the vegetation present at the site. A planned site reconnaissance was realized to describe the type of vegetation in a number of sample points. Previous vegetation studies (e.g. Lundell, 1934; 1937) were consulted in order to identify the plants on a broad level and to generate a map of the vegetation at a scale of 1:20 000. Besides, studies of current vegetation previously completed in the Calakmul Biosphere Reserve (Gutiérrez, 2000) and phytolith analyses of ancient vegetation near Calakmul (Gunn et al., 2002) were valuable for identifying the species. In addition, sample cores for pollen identification were collected in wetlands and reservoirs.

4. To identify the soils of the area and their distribution. The soils of the area were identified in order to describe their physical characteristics and their implicit agricultural and/or functional potential. With the aid of the information given by the planned survey test pits and after subjecting soils to the physical and chemical routine analyses, a soil map of Calakmul at a 1:20 000 scale was generated. The soils of Calakmul Biosphere Reserve have already been identified by FAO scholars on a 1:1 000 000 scale, so a good starting point was already been made (FAO, 1970). Similarly, other soil studies of the area were considered for generating the map (Morales, 1999; Morales and Magaña, 2001).

5. To assess land evaluation. Land evaluation is the process of estimating the potential of land for alternative usages, such as arable farming, grazing, or forestry (Dent and Young, 1981). One of the products of land evaluation is the land suitability classification. The approach for conducting land evaluation is to isolate the relevant land attributes and to divide the range of values or intensities associated with each attribute for considering more than one viable use.

For these purposes, Geographic Information Systems (GIS) were addressed. New technologies make it possible to study the spatial relation between landscape elements, such as the analysis of settlement distributions, spatial patterns, and communication possibilities, among other things. GIS have been used in archaeology for years and they have proven useful in providing a clearer picture in investigations that deal with a high number of variables (Savage, 1990). GIS are digital databases that share a common spatial referenced coordinate system, whose objective is to map and analyze earth features (Davis, 2003). Through them it is possible to introduce, save, transform, and analyze data, as well as generate cartographic products. GIS allow data management through layers and can help to answer complex questions that involve huge datasets of information.

In this work, GIS have been used to generate maps using previously published information and the data obtained in the field by the author. Each landscape element, i.e., relief, soils, vegetation or hydrological features, was analyzed with the aid of GIS software. Each feature was visualized as a layer. To appreciate human remains in the setting, archaeological landscape remnants as roads, structures, artificial reservoirs or channels were handled individually. After doing this, a complex web of relationships was revealed. Although not all elements have the same weight at each point in time, the effort to generate plausible hypothesis in response to specific questions was supplemented. The maps generated were: slope and aspect maps, digital elevation model, orthogonal projections, soil, vegetation, and land suitability maps. Through their combination and analysis it was possible to approach the spatial questions concerned with this study. In this study different software was used. The main programs were Autodesk Map 6 (Autodesk), ArcGIS 8.3 and ArcGIS 3D Analyst Version 8.3 (Economic and Social Research Institute software, also known as ESRI).

6. To identify agricultural and water management features. Although hydraulic modifications for water management, such as *aguadas* (water reservoirs), have been reported before in Calakmul (Domínguez Carrasco and Folan, 1996), it was of great importance for this project to search for other hydraulic features, as well as terrain modifications for agriculture. As a consequence different sectors of the site were explored in order to identify terraces, channels or elevated fields, since they are excellent indicators of the way people modified their landscape.

7. After accomplishing the former objectives, the investigation turned its attention to a discussion of the potential ways in which the land was managed, the plausible use(s) of distinct agricultural methods, and the manner in which people took advantage of land for agricultural, among other themes. The interpretation of the land evaluation potential, the number of inhabitants and the analyzed materials from the test pits were useful topics for suggesting how the ancient Maya modified the landscape.

The Contributions of this Study

Although some studies related to landscape and environment in the Calakmul Biosphere Reserve have previously been undertaken (e.g. Folan and Gallegos, 1998; Gunn et al., 2002), they were done in distant sites, some of them located about 20 km from the archaeological site of Calakmul. Hence, the contribution of this work is to apply landscape archaeology methodology in situ and then to make interpretations based on landscape evidence of the site.

In addition, this investigation could potentially prove useful modern purposes, in terms of suggesting solutions to current environmental problems. Several attempts to manage the rainforest of Calakmul Biosphere Reserve have been realized with the goal of making it suitable for human inhabitation (García Gil and Pat, 2000; Morales and Magaña, 2001). Few of them have been successful, primarily due to the fact that the settlers that actually inhabit the reserve surroundings have modified the place without a deep knowledge of their environment (Morales and Magaña, 2001). Current inhabitants in the Calakmul Biosphere Reserve, most of them from other parts of the Mexican Republic, have impacted the environment negatively, resulting in desertification and soil loss (Morales and Magaña, 2001). Understanding how the Maya developed a culture that thrived and survived for centuries could be of great help in addressing this issue. A basic assumption is that in order to thrive for centuries, the ancient Maya agricultural and water management systems and landscape modifications must have been sustainable in the long term. The observations derived from this study could be taken into account in developing programs for sustainable rainforest management in the region.

Why Archaeological Landscapes?

The purpose of studying ancient landscapes is to visualize, and in some way reconstruct, the societies that inhabited them. The setting, in persistent change as time goes by, is the matrix in which people live. The modes in which subjects respond to the environment reveal much about their customs, adaptations, lifeways, and worldviews. Therefore, landscape analysis is important in that it can reveal clues about how nature has influenced in shaping human lives, and, in turn, about how much people have modified nature in order to fit their lifeways.

Landscape archaeology studies have proven valuable in addressing and seeking appropriate responses to queries regarding the growth of Maya civilization (e.g. Adams et al., 1981; Aliphat, 1994; Fedick and Morrison, 2004; Liendo, 2002). Through the description and interpretation of the physical setting of a given site, they have suggested and posited new ideas regarding subsistence, agriculture and social organization. The physical setting of the cities, their urban context, architectonic functionality, techniques of water management, the methods through which they adapted to environmental conditions and procured basic resources, and their interactions with neighboring societies have all been useful starting points for investigations that seek to construct a vision of particular sites and their environments within an integrative framework.

Along these lines, important aspects of the ancient Maya site of Calakmul and its periphery have been investigated from within a landscape perspective, such as its settlement pattern, ecology, and biodiversity (e.g. Folan et al., 1999; 2001a). With the passage of time, more information has been generated and new technologies have been developed. These new tools allow the integration of extant and recently acquired data in order to refine our comprehension of the ancient Maya and their landscape. Hence, prior research in conjunction with recently recovered evidence showed in this study will provide a further interpretation of the setting and the people that inhabited Calakmul.

What is a Landscape?

"Landscape" is a complex concept and consequently, several definitions of it have been proposed. The discussion of the meaning of landscape across cultures, spatial and temporal contexts, and even individual researchers could be the subject of a complete dissertation. Likewise, many disciplines, such as architecture, geography, history, politics, theology and anthropology, have considered landscapes as important objects of study.

In this study, the landscape is a dynamic place subjected to endogen, exogen and cultural influences without a uniform rate or scale of change. The endogenous factors are those that govern the formation of the Earth's relief and are part of the terrestrial cortex activity, for example, orogenesis or epirogenesis. The exogenous factors are processes related with the atmospheric medium that modify the Earth externally, such as erosion or climate imprints (Lugo, 1988). Although human beings are also exogenous factors, their culture represents a distinctive degree of influence. Culture refers to the learned repertoire of thoughts and actions exhibited by the members of social groups, transmitted from one generation to the next, independently of genetic heredity (Harris, 1979). In sum, endogenous and exogenous factors as well as human culture make each location a unique place on the planet.

As mentioned before, landscapes change without a uniform rate or scale. Physically, some locations can remain in stasis for thousands of years, while others may change several times in a relatively short time. Subsequent generations inscribe their own impressions on a landscape and at the same time remove others left by their predecessors, as in a palimpsest (Ashton and Rowley, 1974). Moreover, some landscape "sections" will change in different manners within a similar spatial context. In addition, the scale of change can be dramatic or minor: consider, for example, the global effects of glaciations versus localized climatic variation.

Studying Landscapes

In order to study a landscape, its two basic levels, physical and cultural, should be approached. The physical setting is formed by the geology, relief, hydrology, climate and living beings (endogenous and exogenous factors); the cultural, by humans' influence and the way they respond to their surroundings. Social schools of thought have usually valued one above the other. For example, positivist archaeologists and geographers emphasize the physical characteristics of the landscape. They focus in technological or economical aspects of culture, and less in social organization, values or beliefs, much more difficult to recognize (Trigger, 1992:304). In contrast, post-processualist scholars highlight the cultural and symbolic aspect of the setting.

But it has been shown that both approaches have the same importance. Knapp and Ashmore (1999:20) affirm "whatever our own traditional views, it is now clear that landscape is neither exclusively natural nor totally cultural: it is a mediation between the two". Likewise, Bell and Walker affirm:

"Economic and social development also had a profound impact on the landscape and within the last five millennia anthropogenic activity in the temperate mid latitude zones has become almost as important as natural agencies in determining the direction and nature of landscape change. However, there is more to the relationship between people and environment than this, for while on the one hand there can be little doubt about the extent of human influence on the natural environment, on the other hand there is compelling evidence to show that landscape and environment have imposed (and continue to impose) their own constraints on human activity" (Bell and Walker, 1992:1).

Hence, both levels should be approached as part of any archaeological investigation.

Visualizing the cultural and natural levels of landscape is just the first step for studying them. In order to get a clearer picture, researchers must go beyond the artifacts and ecofacts per se, and place them in their context. As Butzer emphasizes:

"My plea is for deliberate exploration and development of an approach that will transcend the traditional preoccupation with artifacts and with sites in isolation, to arrive at a realistic appreciation of the environmental matrix and of its potential spatial, economic and social interactions with the subsistence-settlement system. The human ecosystem so defined will open up truly ecological vistas that have been largely neglected. This contextual approach, heavily dependent on archaeobotany, zoo-archaeology, geoarchaeology, and spatial archaeology, is new not in terms of its components but by virtue of its integrated, general goal of understanding the human ecosystem" (Butzer, 1982:12).

Following the contextual approach, the inclusion of geoarchaeological methodology will comprise an important part of this work. The main objective of geoarchaeology is to integrate the landscape through the use of theories and methods derived from the earth sciences (Gladfelter, 1977). Geoarchaeology has two main divisions: the study of the context of the setting, and the consideration of the sedimentary matrix. The palaeo-environmental reconstructions can be based on studies that result from the observation of geomorphology, stratigraphy and sediments in conjunction with soils, fauna, macrobotanical remains and pollen (Shackley, 1981) or so called proxy records. The term proxy is used to refer to any line of evidence that provides an indirect measure of former climates or environments and can include materials as diverse as pollen grains, insect remains, glacial sediments and tree rings, as well as data on crop yields, harvest dates and parish records (Bell and Walker, 1992:11). As Butzer (1982:35) affirms "a competent geo-archaeologist should be able to evaluate diverse sources of empirical data, as generated within the archaeological project and as available from external sources, in order to apply the information to construct an integrated model of a geo-environmental system." Following this school of thought, the study of the landscape, with an emphasis in soils will be significant for approaching the setting of Calakmul.

Landscape Archaeology

Landscape archaeology is a discipline that describes and interprets human interaction with its surroundings from an integrative point of view. At present, it is a very popular branch of archaeology, with several extant studies that follow its theoretical background (e.g. Ashton and Rowley, 1974; Muir, 2000). Landscape archaeology has been defined as a geographical approach whereby a region is investigated in an integrated manner, studying sites and artifacts not in isolation, but as aspects of living societies that once occupied the landscape (Clark et al., 1998).

Landscape archaeology suits different kind of studies and purposes because of its flexible propositions and methodology. Methods from other sciences and sub-disciplines are included considering the basic aims of an investigation. Moreover, new technologies have been incorporated as well as multidisciplinary research (e.g. Pasquinucci and Trément, 2000). This fact lets us join archaeology with "hard" sciences. Impressive discoveries have resulted from this marriage (e.g. Ford, 2003; Sever and Irwin, 2003).

An important facet of landscape investigation is that the specific objectives of any particular study shape the perceived conformation of a given landscape. On one hand, a landscape could be envisioned as a garden or a plaza (Miller and Gleason, 1994), considering features on a grand scale, a fact that contrasts with the traditional archaeological definition of the site. On the other, the scale of a site can be small, interpreting features from the surface (Erickson, 2001) or defining clusters of sites within large regions.

Landscape Researchers

In the Western world, European scholars made the first organized attempts to study ancient landscapes in the nineteenth century. In 1840, the director of the National Museum of Denmark, J. A. A. Worsaae, discussed the importance of considering archaeological remains in the setting in which they were found (Trigger, 1992:235). Later, twentieth century scholars as Sauer, Clark and Bradford set the bases for landscape studies.

Carl Sauer influenced the emerging North American landscape school. In his study The Morphology of Landscape (Sauer, 1925) he discussed three main questions: What is the relevance of landscape? Why do human groups evolve in different ways? How have humans modified their landscape? For this "landscape school", culture is the physical evidence of human activity in a given area. So the primary goal is to study the cultural landscape, not just the culture (Norton, 1989:36-39). Sauer pointed out that a firm knowledge and understanding of historical sources, archaeological sites, biogeography, ecology, and the processes of geomorphology must be fused with field studies, so that we may read and interpret the changes in habitability through human time for the lands in which civilization first took form (Sauer, 1940).

The British archaeologist Grahame Clark, a pioneer of ecological archaeology, proposed that an ancient settlement is related to the climate, vegetation and fauna of its own spatial context. He emphasized that geographic conditions should be traced in each period of the human settlement, and that it is also important to recognize the environmental limitations, in order to determine which economic activities can be reflected in the landscape, and what can be reconstructed with the aid of the geographic features (Clark, 1964).

Bradford was one of the first researchers to interpret aerial photographs in landscape studies by observing and classifying the setting elements (Bradford, 1957). He demonstrated that aerial photographs are valuable as objects of study in and of themselves, and not just as auxiliary elements useful in site description. As a consequence, they have become a useful resource for archaeologists in a wide variety of projects.

In the late 1950s, as part of the emerging 'New Archaeology' school, archaeologists focused on the ecology of the ancient settlements, emphasizing technological aspects over social organization, values, or beliefs (Trigger, 1992:304). Lewis Binford, one of the founders of New Archaeology, posited that biological sciences group more objective data, and that these kinds of data can elucidate how processes affect the setting. As he states, "the idea that specific processes may operate in the present as well as in the past was the central formulation response of the sciences of geology and paleontology, two related disciplines that also deal with the dilemma of knowing the past and the present" (Binford, 2001:10).

In a similar vein, the central tenets of the 'New Geography' state that there are spatial and morphological laws that can be identified with a strict methodology following the scientific method (Wagstaff, 1987). This tradition has influenced leading researchers who have studied the landscape change in the New World, particularly in the Maya area (e.g., Dunning et al., 2002; Turner, 1983).

During the 1980s, the post-processual school emerged as a reaction against the positivist approach of New Archaeology. They directed attention toward elements such as identity, perception and cultural meaning in cognitive and/or psychological terms. The foremost representative is Ian Hodder. With respect to landscape, he affirms, "the relationship between material culture and human organization is partly social...but it is also dependent on a set of cultural attitudes which cannot be predicted from or reduced to an environment" (Hodder and Hutson, 2003:4).

In the last few years many subdivisions and branches of distinct disciplines have been derived in order to study the concept of landscape that differ from the framework and methodology of landscape archaeology. As a consequence, we encounter studies with a phenomenological perspective (Koontz et al., 2001; Tilley, 1994), a historical ecological view (Crumley, 1987; Erickson, 2003) or the landscape approach (Rossignol and Wandsnider, 1992).

Landscape Archaeology in the Maya Area

Mayanists have long been interested in environmental changes, features of the physical environment, and landscape management. Previous studies can be divided in three broad areas. The first is related to the multidisciplinary approach. Biological sciences methods applied to some archaeological questions have confirmed previous hypotheses related to the Maya. For example, scholars have detected the presence of a dry period with erosion and deforestation indicators that correlated significantly with the Maya collapse (e.g., Curtis et al. 1996, 1998; Hodell et al., 1995; Islebe et al., 1996; Rosenmeier et al., 2002).

The second branch reunites scholars who analyze the cultural and symbolic element of the landscape (e.g., Koontz et al., 2001). Such studies ruminate on the ways in which the Maya perceived and interpreted landscapes, focusing on themes related to culture, symbolism, ritual or religion.

The last division groups related research with historical ecology, landscape archaeology, and geoarchaeology. Although researchers also employ the newest methods of the so-called hard sciences, their focus is more based on a purely archaeological methodology. Hence, topics such as geography, subsistence, agricultural systems or settlement are explored (e.g. Dunning et al., 1999; Fedick and Morrison, 2004; Harrison, 1996). For example, in a study of the upper Usumacinta River archaeological sites, natural landscape information was transformed into maps with the purpose of generating a political landscape of Yaxchilán and Piedras Negras during the Classic Period, relating at the same time the physical and cultural features of landscape (Aliphat, 1994). This dissertation is related with this branch of Maya landscape archaeology.

In some cases, the boundaries between each of the three identified branches are not easily distinguishable. Each, however, has undoubtedly contributed to the way we think about the Maya.

The Way of Approaching Landscape in this Study

The purpose of this study is to identify and collate ancient Maya landscape elements in order to emphasize the potential resources and advantages that gave rise to the development of Calakmul. The first assumption is that the environmental conditions changed persistently throughout the Holocene. Hence, this state of continuous transformation is considered as a constant for Maya development. As Stahl affirms, in the Americas "at any given time or place, both the distribution of the numbers of different kinds of organisms and their relative abundances were in a constant state of flux since the retreat of glacial cover some 10 000 years ago" (Stahl, 1996:105). The climate in Yucatan Peninsula, as is the rule, is not stable. It presents droughts and periods of excess rain that endure for variable time spans. So, when analyzing the Maya of Calakmul, the only constant that will be taken in account is the fact that their environment continually changed during the time they inhabited the rainforest.

But certainly, this change implicates a threshold. The presence of rainforest fauna such as tapirs, monkeys or jaguars, as seen in ancient inscriptions, evidences that the Maya observed these species in the surroundings. The same animals were seen in the Maya area during the sixteenth century (de Landa, 1994) and at present they subsist in Calakmul Reserve. Hence, it is possible to suggest that a very similar, although not equal, natural setting would have existed in the past. As population increased, wild areas became reduced significantly.

The environment characteristics are not to be taken as deterministic factors that necessarily affect societies in a predictable way. But they indeed have a big influence in the way in which people adapt to their setting. A big emphasis will be given to the strategies that the Maya posed for adapting to their environment Each strategy, as reflected in the archaeological record, will be an indicator of institutions or ideology that allowed an adjustment to the landscape (Bennet, 1976).

Adaptation is defined as the anatomical, physiological, or behavioral trait that contributes to an individual's ability to survive and reproduce in competition with conspecifics in the environment in which it evolved (Williams, 1966). Through this behavior humans seize opportunities and resources in their environment.

Human strategies of adaptation will be figured as retroactive responses to their environment. On explaining cultural changes, Erickson affirms, "an archaeology of landscapes, combined with the insights of the New Ecology, provides a theoretically robust alternative. In this perspective, humans are considered active and dynamic agents who not only respond to the challenges of fluctuation of climatic in their environments, but also create, shape and transform those very environments" (Erickson, 1999:641).

The next assumption is that the Maya reacted to environmental elements in varied manners in order to subsist. Thus, great diversity existed in adaptive responses. As Fedick points out "the tropical lowlands of southern Mexico and Central America are composed of a complex mosaic of landscapes that presented a variety of possibilities and challenges to the subsistence practices of the ancient Maya" (Fedick and Morrison, 2004:207).

In the same way, Lentz approaches this diversity through the understanding of the ecology of the New World populations, presenting their involvement in local ecosystems and focusing on their interrelations with the environment. He concludes that in order to survive ancient New World societies responded with varied actions to their setting, with sustainable and highly productive land use practices (Lentz, 2000).

The approach employed in this study considers that a landscape is the result of its particular geology, geomorphology, soils, climate, hydrology and biodiversity (that is the represented by all living organisms, flora, fauna and human beings), all of them interrelated as in a machine (Zonneveld, 1979). In this model, each feature represents a gear, subjected to constant change, affecting all the other gears of the system and forming altogether the landscape (See Figure 1.1). Humans play an active role in this interrelation. Cultural expressions affect the setting in a tangible way.

In this model, it is indispensable to examine the relationships of each landscape feature in order to establish minimal units of analysis (landscape units). If for example, Calakmul Reserve rainforest is only observed through remote sensing images, large homogenous areas become visible. But in reality, the setting varies. This variation is explained by each element's characteristics (each gear) and its interaction with all the others.

Although this model for studying landscape has a reductionist slant, it is important to note that in the end, the setting should be considered to have an emergent quality, since its complex whole is much more than the simple sum of its component parts. Through archaeological and ecological methodology, studying the parts is a useful starting point, practical for considering large quantities of variables and their interactions.

After analyzing the landscape features, and consequently, the landscape units, the conditions of Maya landscape – how did they develop strategies for living in it and modifying it– will be proposed.

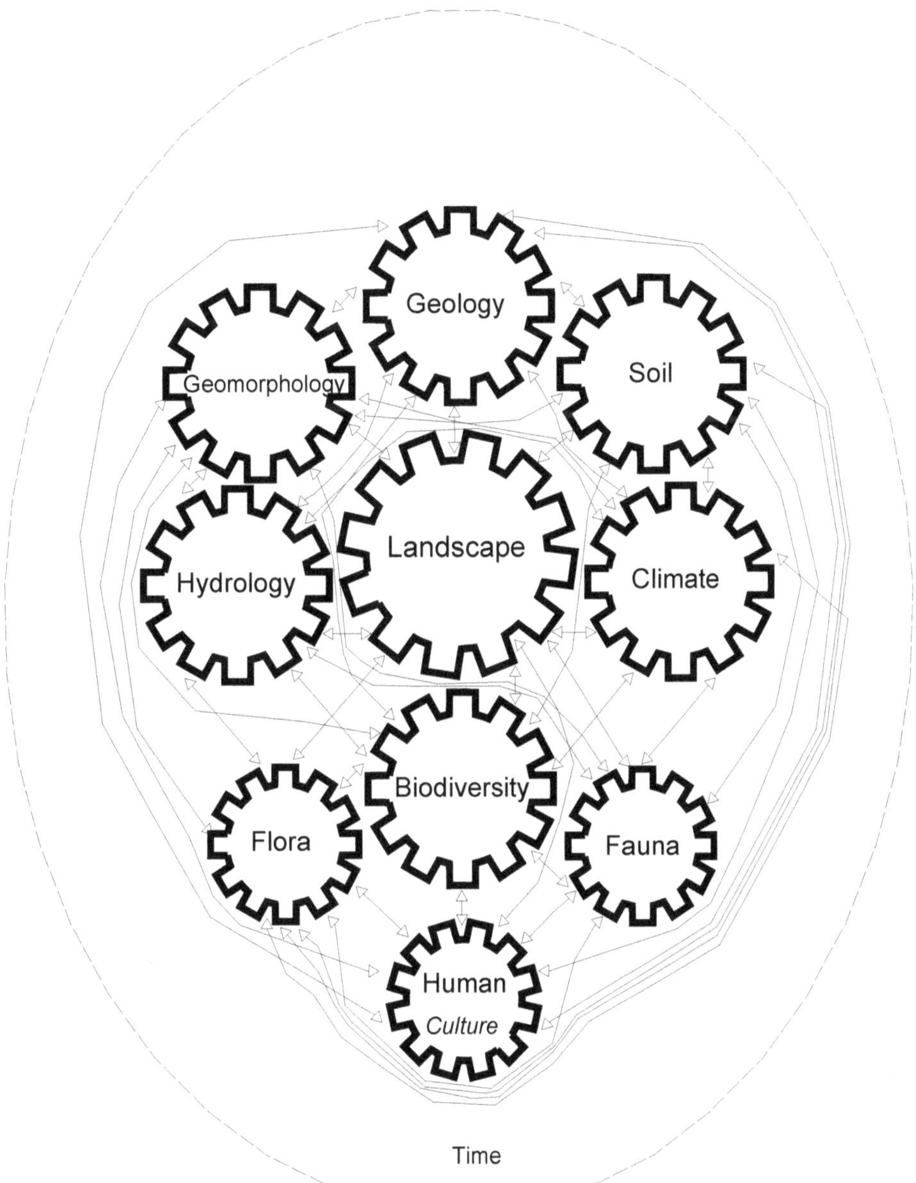

Figure 1.2. Model of landscape dynamics. Each landscape feature (represented as a gear) is considered to be in constant change and interacting with the others. This model is based on Zonneveld (Zonnenveld, 1979).

CHAPTER 2. THE MAYA KINGDOM OF CALAKMUL

"At dawn the next morning, the twenty-ninth of December, we headed for these ruins, going in a southwesterly direction, following a narrow muddy chiclero trail. We soon saw many signs of former habitations. After riding about seven miles, we came to rectangular and pyramidal mounds. At half past three in the afternoon I saw, standing in the trail before me, a huge sculptured monolith. We passed it, and a pyramid to our right, and entered what I found to be the main plaza and the ceremonial center of one of the most important cities of the Southern Maya Culture."

Cyrus L. Lundell describing his entrance to Calakmul in Lundell, 1933:152-153.

Mesoamerica

The Maya site of Calakmul is located in a region that has been called Mesoamerica or Middle America (See Figure 2.1). Mesoamerica is a concept proposed by Kirchoff that denotes the sharing of particular cultural traits in a specific time and place (Kirchoff, 1943). This concept was created in part by considering 16th century inhabitants who shared cultural features, involving the present-day countries of Belize, Guatemala and El Salvador and part of Mexico and Honduras. Some of the shared cultural traits are the construction of stepped pyramids, the painting on bark paper codices, the practice of human sacrifice, the use of *coa* (digging stick) for planting and similar calendar and writing systems, as opposed to their North-American and South-American neighbors.

The Mesoamerican cultural zone has been subdivided into different areas that are: Northwest, Central Plateau, Gulf Coast, Oaxaca and Maya. Each one is characterized by distinctive regional expressions of the mentioned cultural traits. Stepped pyramids, for example, can be appreciated in each one of the areas, but certainly, a Maya pyramid looks different from one of the Gulf Coast.

Maya Area

The Maya Area comprises approximately 600 000 km^2 with a perimeter of about 5500 kilometers. It has been subdivided according to physiographic and cultural shared traits. Several subdivisions of the Maya Area have been proposed and the broader one was chosen for simplicity matters. The general Maya Area subdivisions are: Northern Lowlands, Southern Lowlands, Guatemala and Chiapas Highlands and Pacific Coast (See Figure 1.1). The term lowland depicts the almost plain surface of the terrain in contrast with the highlands. The Northern Lowlands area covers almost all of the three Mexican states of Quintana Roo, Yucatan and Campeche. Northern Yucatan is characterized for its almost flat karstic platform, lacking surface rivers and having a low and medium rainforest. Sites such as Chichen Itza, Uxmal, Xcambo and Mayapan flourished in the north of the Yucatan Peninsula. The Southern Lowlands comprise southern Campeche and Quintana Roo, portions of eastern Tabasco and Chiapas Mexican states as well as the Guatemala Department of Petén, Belize, western Honduras and El Salvador. The portion lying south of Lake Petén Itza is a tropical humid forest; to the north, the terrain is covered with tropical dry forest. Swampy areas or *bajos* cover the ground with few permanent streams. Sites like Tikal, Palenque, Uaxactun or Calakmul represent some sites that thrived in this area. The Guatemala and Chiapas Highlands as well as a significant area around Huehuetenango, Guatemala, and western Salvador are covered with lower montane and montane pine and oak humid forests. Sites as Takalik Abaj, El Baúl, and Kaminaljuyu are few examples of the sites that prospered in this region. The Pacific Coast is an area in which early societies developed, such as Izapa, Cerro de las Conchas, Tlacuachero or Manchón (Neff et al., 2001).

Mesoamerica and Maya People through Time

The most accepted model regarding the first settlers of America has been the one that proposes that they came from Asia via the Bering Strait land bridge about 11 500 years ago descending gradually from North to South America (Hoffecker et al., 1993). These groups continually migrated in keeping with sustenance priorities and consequently populated the American continent. Stones, points and transition camps are the remnants left by the first American inhabitants. However, early dates of 11 500 B.C. in Panama (Piperno and Pearsal, 1998) or 10 500 B.C. in Monte Verde, Chile, challenge this hypothesis (Gibbons, 1997). A better explanation must be found in order to clarify how people reach the south in such early dates. This debate continues until today and it is expected that new research will unveil this transcendent topic.

It has been difficult to trace the way of life of the first inhabitants of the Maya area. The first reason is that the tropical environment is not favorable for the conservation of organic remains. The second reason is that few researchers have been interested in retrieving this kind of information. Fortunately, this last situation is changing and probably in the future we will have a better picture of the earliest Maya.

In Mesoamerica, during the Lithic Stage Arqueolithic period (34 000/33 000 B.C. to 9500 B.C) nomad groups modified lithics in a rudimentary way. Blades, scrapers and flakes were collected from Tlapacoya, Estado de México and El Cedral, San Luis Potosí (Mirambell, 2001). At Tlapacoya blade-like flakes were found near a hearth dated 18 000 B.C. (Lorenzo and Mirambell, 1986).

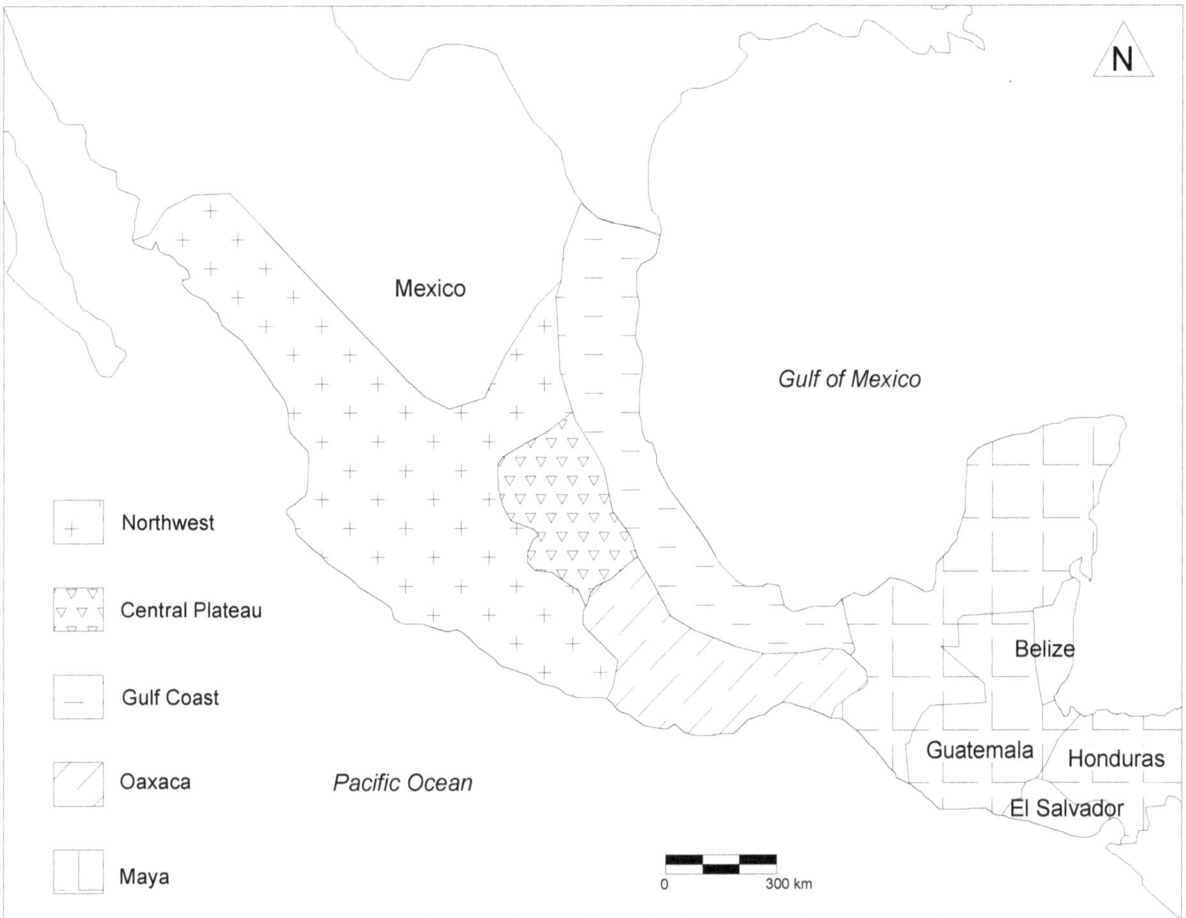

Figure 2.1. Mesoamerica and its subdivision.

As a result of the Pleistocene-Holocene transition, major climatic changes as global warming occurred on the Earth. Hence, glaciers receded at upper latitudes in conjunction with the rise of the sea levels. As a consequence, Pleistocene megafauna were extinguished and the landscape was dramatically modified. During the Lithic Stage Inferior Cenolithic Period (9500 B.C. to 7000 B.C.) environmental circumstances triggered new strategies for adaptation. Hence, artifact complexity increased as seen on tools such as Clovis points in Rancho Weiker, Durango or Teopisca, Chiapas, as well as "fishtail" points found in Los Grifos Chiapas (García Barcena, 1979; Mirambell, 2001). At Los Tapiales, Guatemala, the base of a Clovis point, bifaces, burins, scrapers, and blades dated 9600 B.C. and 8800 B.C. have been found (Gruhn and Bryan, 1977). The Cave of Santa Marta in Chiapas and El Gigante in Honduras have provided a picture about the way Lithic Stage inhabitants exploited their environment (Scheffler, 2001).

Throughout the Lithic Stage Superior Cenolithic Period/Protoneolithic (7000 B.C. to 2500 B.C.) stones were polished and axes and grinding stones were made (García Bárcena, 2001). In addition, the origin of agriculture was a transcendent event of this period. Sites as Cueva La Perra and Presa Falcón in Tamaulipas and Tehuacán, Puebla represent this period (MacNeish, 1964). During this lapse of time, nomad hunter-gatherers groups dispersed across large areas, living in seasonal camps, collecting, domesticating and processing plants as well as taking advantage of marine resources. They had a rough life in order to adapt to external conditions. They began to establish the first permanent settlements and to domesticate and cultivate species as maize, squash, chili, avocado, beans, *guaje*, plum, *zapote* and amaranth (McClung, 1993).

Since the beginning of the Formative or Preclassic Period, Mesoamerican people changed their way of living: the first villages were developed, agricultural techniques were improved for the growing, increasingly stratified population and the earliest pottery was made. Ceramics found in different sites of contemporary Guerrero, Oaxaca, Puebla, Jalisco and Colima states, as well as in Chiapas and Guatemala Coast account for this change (Serra, 1993). The first village dwellings, made with perishable materials, gave shelter to egalitarian inhabitants who buried their relatives below their homes.

The Preclassic Period has been divided in Early, Middle and Late, and it lasts from 2500 B.C to A.D. 150/200 (López Austin and López Luján, 2000).

The first signs of sedentary life in the Maya Lowlands are noticed during the Early Preclassic Period, which begins in 2500 B.C. and ends in 1200 B.C. Lithic scatters with patinated flakes, projectile points, sandstone bowls, unifaces and large blades have been found in sites with aceramic contexts north to Belize City, with radiocarbon dates to ca. 2500 B.C. to 1500. B.C. (Hester et al., 1995). Archaeological remains indicate that the first settlers of Copan occupied the valley about 2000 B.C. (Fash, 1991). Preference for estuarine or riparian settings, timber-framed houses, enhanced use of lithics and pottery are just some of the features of this period. At Paso la Amada in Chiapas, the first buildings for public use are seen (Ceja Tenorio, 1985). The ceramics of the phases Barra, Locona and Ocos respectively, are observed in the Guatemalan Pacific Coast (Coe, 2000a).

The Middle Preclassic period, from 1200 B.C. to 400 B.C, is characterized by the influence of Olmec culture and the boost of agricultural villages in Mesoamerica. During this period large-scale terrain modification with urban and agricultural purposes is seen for example at La Venta, and in Takalik Abaj. The site of Cuello, Belize, has been an excellent example of the beginnings of the sedentary life with an occupation 1200 B.C. to A.D. 400. It has been possible to observe the architectonical evolution of timber and palm with clay floor houses to apsidal dwellings with *bajareque* walls and lime floor built above platforms (Hammond, 2001). In sites as Nakbe, households with organic materials (wood poles, thatch) built above platforms are seen. Major developments as monumental architecture and city planning are also evident. By the end of this period pyramids of up to 18 meters high have been identified in Nakbe, as well as platforms that covered about 40 000 m^2, Group E astronomical complexes and a ball court (Hansen, 2001). In this period the first monuments with writing also appear. Sites as Tintal, Nakbe, Isla, Pedernal and possibly El Mirador expose the first stelae and altars (Hansen, 2001). This time marks the beginning of early state-level societies in Mesoamerica. An interesting feature is the elaboration of objects with shell in sites like Nakbe, Uaxactun, Tikal, Cahal Pech and in the Río Pasión region (Hansen, 2001). Many Maya sites of this period have been studied such as Komchen, Kaxob, Tikal, Uaxactun, Nakbe, Lamanai, Seibal, Altar de Sacrificios, Yaxuna, Cahal Pech, Chalchuapa area and Calakmul.

The Late Preclassic Period, from 400 B.C. to A.D. 150/200 is marked by the collapse of La Venta and major Olmec influence. Many small settlements developed into bigger centers in the Central Plateau of Mexico, the Puebla-Tlaxcala basin, Oaxaca and Maya Lowlands. This period can be seen as the first apogee of Maya culture. Triadic pyramids became visible. Huge pyramids up to 40 to 72 meters high were constructed in sites as El Mirador, Nakbe, Tintal, Wakna, Xulnal, Paixban, La Manteca and Naachtun (Hansen, 2004). Some structures of Yaxha, Lamanai, Cerros, El Mirador, Edzna, Chiapa de Corzo, El Tigre and Calakmul were decorated with giant gods stucco masks. *Sacbeob* or Maya roads were built connecting city plazas and even other sites. The representative ceramics of this period in the Maya area is Chicanel (Coe 2000a). Elaborated funerary precincts indicate social ranking. Sites like El Portón, el Baúl, and Takalik Abaj show monuments with hieroglyphic texts. Fortifications were built in some sites such as Becan, Cerros, Edzna and El Mirador. This is a sign of conflict and could give us an idea of wars or revolts that appear to have promoted the abandonment of some centers. Many settlements as El Mirador, Uaxactun, Seibal, Cerros, Colha, Becan, Dzibilchaltun, Komchen, Isla Cancun, Edzna and Santa Rosa Xtampak were abandoned during the transition between the Preclassic and Classic periods, and the reason is still not clear (Hansen, 2001).

Throughout the Classic Period, the complexity of almost all cultural expressions is manifested. Writing monuments with a well-developed system, and exchanging goods through long distances was a landmark of this epoch (Manzanilla, 1993a). Goods as jadeite, flint, obsidian, shells, cotton, salt, copal, feathers or cacao were some of the most sought-after merchandise in Mesoamerican, traded over very long distances. During this period the Northern Lowland people spoke the Yukatek Mayan language, the Southern Lowland inhabitants, Cholan Mayan and the Highland and Pacific Piedmont inhabitants spoke Southern Maya (languages which includes Greater Quichean, Mamean, Greater Kanjolabal and Tzeltalan-Cholan; Coe, 2000a). The Classic Period is placed between A.D. 150/200 to A.D. 900 (López Austin and López Luján, 2000). During the Classic Period, the emergence of the Teotihuacan dominion, the power of major Lowland Maya sites as Calakmul or Tikal, and the authority of the Zapotec kingdoms of Monte Alban developed.

The Classic Period has been divided in Early Classic, from A.D. 150/200 to A.D. 600, when Teotihuacan thrived, and Late Classic from A.D. 600 to A.D. 900, when Lowland Maya experienced their second apogee. Some scholars also consider transition periods as the Protoclassic period from A.D. 100 to 200, the Terminal Classic period that lasted from A.D. 900 to A.D. 1000 as well as an Epiclassic period from A.D. 700 to A.D. 900 (Sugiura, 1993) in the Central Mexican Basin (basically applied in Mexico's Central Plateau after the disintegration of Teotihuacan).

The Early Classic Period is characterized by Teotihuacan supremacy over much of Mesoamerica (Manzanilla, 1993b). The city core was built following a careful astronomical plan and its monumental architecture still stands. Its influence reached distant places in the Lowland and Highland Maya area, such as Tikal or

Kaminaljuyu. Recent studies support that Teotihuacan and Maya sites interacted in a bidirectional way, and that this relationship changed within each site (Braswell, 2003). Besides Monte Alban, with its particular writing system and architecture dominated the Oaxaca Valley during this lapse of time.

Kingdoms such as Tikal, Uaxactun, Calakmul, Yaxha, Xultun and Naachtun that developed during the Preclassic, increased their size in the Classic in contrast with other sites of the Mirador Basin, which were abandoned. During the Early Classic the Maya gained in complexity as can be evidenced by the increasing number of sites, architectural achievements as the corbelled vaults, plazas or monumental inscriptions with dynastic records. Later dynasties emerged in sites like Copan, Yaxchilan, Calakmul, Tikal or Palenque with the king *ajaw* being the inherited ruler and representing the direct connection with divinities (*k' uhul ajaw*; Grube and Martin, 2001). Since the last century scholars have debated where Classic Maya developed into many independent minor states or few big potencies or "superstates" (Martin and Grube, 1995). What has been shown by emblem glyphs is that there were numerous dynamic kingdoms whose power did not always last for long.

During Late Classic Period, the power of Teotihuacan and Monte Alban ceased (c.a. A.D. 600; Sugiura, 1993). As a consequence, many medium centers as Xochicalco, Tajín, Magoni, Cacaxtla or Cantona appeared in order to fill the lack of power. It seems that war was almost constant during this period. A word that has been employed to describe the situation that prevailed during the Late Classic has been balkanization (Marcus, 2001:23). Balkanization means that small states fight between each other and then form bigger states, but with an unstable political sphere. This same phenomenon occurred in the Maya area with an increased warfare period that only ceased toward the end of Late Classic, with the so-called Maya Collapse. The existence of alliances, the paying of tribute and the mention of defeated enemies were some indicators of social instability.

During the Late Classic Period, the Maya reached their peak. Multiple sites thrived with large populations. An increase of carved monuments, agricultural intensive techniques such as raised fields, the demand of painted Maya vases, the construction of pyramids for the Maya *ajaw*, the dynastic rites of succession and warfare proliferated during this lapse of time. In the Mirador Basin, corbelled-vault buildings and elite tomb constructions were built on top of Preclassic ruins (Hansen, 2001). Many other Maya sites followed this pattern.

Toward the end of the Classic warfare increased, carved monuments stopped and buildings were not longer remodeled. In Dos Pilas, Aguacateca and Punta de Chimino extensive fortifications were hurriedly erected (Coe, 2000a). As mentioned before, big states became *cuchcabalob* or Balkanized states. The big centers as Calakmul of Tikal had temporary dominion over other minor states, forming alliances. During this 'Collapse', people abandoned the majority of the cities of the southern lowlands (Gill, 2001). The proposed causes of the Collapse have been revolts, invasions and war, a decomposed environment or droughts. However, this was the time when other cities arose such as Sayil, Labna, Seibal, Lamanai, and Chichen Itza. This confronts the total Collapse hypothesis that states that in a much reduced period of time almost all the Maya disappeared.

In the Postclassic Period, the southern Maya cities were already abandoned and new settlements were founded. The Postclassic Period has been divided in Early Postclassic from A.D. 900 to A.D. 1200 and Late Postclassic that lasted until A.D. 1520 (López Austin and López Luján, 2000). Elsewhere in Mesoamerica Tula was founded, Mixtec people gained influence and Oaxaca sites suffered a strong balkanization (Coe, 2000b). Several codices have been documented from this period. Multifaceted military affairs stand out as a principal issue in documents, architecture and sculpture.

During the Late Postclassic, the walled city of Mayapan in Northern Lowlands, the architecture of which resembled Chichen Itza, was the capital of the most centralized state of the this period (Masson et al., 2002). Sites like Tulum, Tancah, Santa Rita Corozal, San Gervasio and Isla Mujeres flourished in the northern lowlands, and Utlatlan and Iximche in the southern highlands (Carmack, 1981).

During the Late Postclassic period, in Central Mexico the Aztecs formed an empire with its capital at Tenochtitlan that ended in A.D 1521 with the Spanish conquest. In the same way Tarascan people, who settled in Tzintzuntan in west Mexico, developed a large state. Products from longer distances were exchanged through *pochtecas*, a special merchant class dedicated to trade products from one region to another. The importance of the use of metals for utilitarian purposes and adornment increased. War activities developed with military expansion, tributes, sacrificial victims, special gods and defensive fortifications (López Luján, 1993).

It is important to emphasize that not all the southern lowland Maya people disappeared in the Late Postclassic period. Topoxte and Tayasal communities resisted the Spanish invasion until seventeen century. The last ancient Maya sites in the northern lowlands were small cities like Tulum, on the Caribbean coast.

The Spanish conquerors were impressed by the Maya civilization they found. With their arrival, the history of a new culture began. Many years later, the first European explorers crossed the region exposing marvelous vestiges found in almost every place of the New World. During the last century, scholars have gained insights of the material culture through the decipherment of Maya writing, the comprehension of the ancient knowledge of astronomy, mathematics and architecture. They have approached the Mesoamerican way of living, especially that of the elite. At present, Maya descendants still live in Mexico and Central America. In some cases they still talk in Maya language and also practice ancient Maya traditions.

The Kingdom of Calakmul

Located to the south of the Mexican state of Campeche, on a limestone promontory next to an enormous *bajo*, stands the site of Calakmul. For centuries this site was dormant and the access was restricted because its inaccessibility. Rainforest toxic plant species, dangerous fauna and muddy swamps impeded human travel through the region. Many long days to reach the site, lack of water, the threat of bandits and worry of vehicle breakdown in the middle of nowhere, all confirm the oral tradition of the site's history.

Nevertheless, looted structures, *chicleros'* camps and even scars in trees, testify that the area had not been completely isolated. The exploitation of *tinta* or logwood (*Haematoxylum campechianum*), a tree prized for red and purple dyes during the last decades of nineteenth century, was an activity that impacted the area (Acopa and Boege, 1998). In the same way, the harvesting of chewing gum sap from *chicle* trees (*Manilkara zapota*) in the late nineteenth century promoted forest exploitation. Moreover, large-scale timber extraction in the early twentieth century contributed to the disappearance of precious wood trees.

During the 1980s the first road was built into Calakmul. As a consequence, a long-term archaeological research project began and many other smaller Maya sites were discovered. The building of a camp and the creation of human resources for maintaining the archaeological site attracted more people. Years later, once the site was declared as protected area in 1989, access for many other people (such as tourists) was permitted to this huge biosphere reserve.

- The Name of the Site

The botanist and explorer Cyrus Lundell gave the current name of Calakmul in the 1930s. Although it is properly written as Kalak'mul, it will be cited as Lundell did, since the great majority of the references allude to the site in this way. The identification of the names related with the kingdom of Calakmul was done through the reading of the emblem glyphs. As Heinrich Berlin proposed in late 1950s, emblem glyphs consist of a variable main sign attached to two unvarying smaller elements (Berlin, 1958). The first of these two elements can be read as *k'uhul* or divine and the second, as *ajaw* or ruler. The third part names a Maya kingdom. As a whole, emblem glyphs function as titles of Maya kings describing each as the divine lord of the kingdom whose name appears as the main sign of the glyph (Martin and Grube, 1995:42). Four names have been associated with Calakmul: The emblem glyph Kan (or *Kaan)* that bears a snake head (Figure 2.2), the emblem glyph Bat (Figure 2.3), and the toponyms *Uxte'tuun* or 'Three Stones" and *Chiik Nahb* (Figure 2.4) of uncertain meaning, probably referring to a water lily plant (Martin, 1996; 2005).

Joyce Marcus was the first to suggest that the *Kan* glyph was related with Calakmul (Marcus, 1973). Years later, Stuart and Houston found an association between the place glyphs *Oxte'tun* and *Nab' Tunich* and suggested that they were related with the Snakehead Kingdom (Stuart and Houston, 1994). Later, the epigrapher Simon Martin pointed out that the name *Nab' Tunich* was incorrect, and suggested *Chiik Nahb* (Schuster, 1997).

Figure 2.2. The glyph of Kan representing a snake's head (redrawn from Martin and Grube, 2002:101).

Figure 2.3. The glyph of the Bat. Stella 62 of Calakmul (redrawn from Martin, 2005:5).

Figure 2.4. Glyphs of Uxte'tuun and Chiik Nahb (Redrawn from Martin and Grube, 2002:104).

The snake glyph has been found in several inscriptions in Lowland Maya area; in fact it is the most mentioned site in the Maya region (Martin, 1996). It has been found for example at the site of La Muerta in the Mirador Basin (Suyuc et al., n.d.), at Tikal and Caracol (Martin and Grube, 2002) and at Dzibanche (Velasquez, 2004). Factors such as site's dimension and antiquity have supported the hypothesis that the snake glyph coincides with the Kingdom of Calakmul (*cuchcabal*) during all Classic period (Carrasco and Colón, 2005). However, the existence of another emblem glyph, the bat glyph, challenges this hypothesis.

The bat glyph has been observed in Calakmul on stelae 114 (A.D. 435), 62 (A.D. 751) and 59 (A.D. 741), and also at Oxpemul, before and after the reign of the three most remarkable Calakmul kings: *Yich'aak K'ahk'*, *Yuknoom Ch'een II* and *Yuknoom Took' K'awiil*, who ruled during the seventh century and beginnings of eighth. This fact has suggested the presence of the Snake Kingdom in Calakmul only during the reign of these kings (Martin, 2005). Besides, the presence of the snake glyph in Dzibanche (Velásquez, 2004) supports this hypothesis and the possibility of different seats for the Snake capital.

The meaning of the other two names of Calakmul, *Uxte'tuun* and *Chiik Nahb*, is still not clear. In some monuments both names appear together forming a toponym which implies that one place is part of another. The glyph *Chiik Nahb* appears in Naranjo, Dos Pilas, Cancuen, La Corona, Tonina and Quirigua inscriptions. In Calakmul, it has been identified on Stela 51, in the Annex of Structure XIII, on an Early Classic plate from Structure IV-B and at Northern Acropolis (called Chik Nab; Carrasco and Bojalil, 2005). Epigrapher Simon Martin proposes that *Chiik Nahb* corresponded to a locality, bigger than *Uxte'tuun* (Martin, 1996:44). Carrasco and colleagues affirm that it denotes a specific urban space: the Acropolis of Chik Nab (Vázquez López, 2006).

It has been proposed that *Uxte'tuun* was the ancient name of the kingdom. The significant title *Uxte'tuun Kalomte* or Three Stone *Batab* accounts for the ritual importance of this place for the Maya (Martin, 1996:44). Reents Budet suggests a cosmogonic interpretation of *Uxte'tuun*. The three stones symbolize the three stones of the cosmic home and similarly they refer to the three stone thrones, where *Hun Hunhahpu* god supervised the creation. Following this interpretation, Calakmul could be considered also as the place where the universe originated (Reents Budet, 1998 in Boucher and Palomo, 2000).

Another interpretation of the name of *Uxte'tuun* could be related with the characteristics of its geography. The city of Calakmul is located in a rounded karstic promontory surrounded by the *Bajo* El Laberinto. This prominence is dissected by three deep ravines positioned to the northwest and south that form a tri-lobed relief (Figure 2.5). Each one of them could represent a stone, and together these tree stones could give to Calakmul its ancient name.

Tuun, the Maya word for stone, is used in a number of toponyms, such as *Lacantun* or 'Big Stone'. Other Maya cities took their names from characteristic geographical features. As Stuart points out "at Palenque's Temple XIX we find *tahn ch'een Lakamha'* ("in front of the Lakamha' spring") probably in reference to the Otolum River (Stuart, 2000). Stela 3 from Caracol mentions the local place name *Uxwitza'* ("Three Hill Water") in much the same way" (Stuart 2004:1). So, it can also be a possibility that the geographical features of the landscape, the tri-lobed promontory in this case, could be reason for denominating the city of Calakmul as *Uxte' tuun.*

- Site Exploration

The first occidental researcher who reported the site of Calakmul was the botanist Cyrus L. Lundell in 1931. He named the site Calakmul, being in Maya *ca* two, *lak* adjacent and *mul* artificial elevations or "City of the Two Adjacent Pyramids" (Lundell, 1933). He and his team also mapped its ceremonial center and identified also some stelae (See Figure 2.6).

In this same year, Lundell went to Chichen Itza and reported the discovery to the archaeologist Sylvanus Morley, who organized the first formal trip to the ruins: The First Campeche Expedition of the Carnegie Institution of Washington, held in 1932. Their members explored the site and registered multiple sculpted monuments (Morley, 1933). In 1933, Enrique Juan Palacios, from Mexican *Dirección de Arqueología,* made a reconnaissance trip (Palacios, 1933). In 1934 Ruppert, Denison and O'Neill visited again the site as part of the Second Campeche Expedition (Ruppert and Denison, 1943). As a result, an accurate map that is still used today was generated (Figure 2.7). Also in the 1930s, Gustav Stromsvik (1937) studied the *metates* (grinding implements) of Calakmul.

In 1975 Eric von Euw visited the site in order to register sculpted monuments for the Corpus of Maya Inscriptions of the Peabody Museum (Carrasco et al., 1996:7). In 1981 the Mexican *Instituto Nacional de Antropología e Historia* commissioned archaeologist Peter Schmidt to carry out the first delimitation of the archaeological zone (Carrasco et al., 1996:7).

In the 1980s, the archaeologists Román Piña Chan and William Folan, as well as a multidisciplinary team, carried out a project with cultural and ecological interests (Folan et al., 1995). They worked in the site from 1982 through 1992. From 1982 to 1989 they mapped 30 km^2 of the site (Folan et al., 2001a; see Figure 2.8). This team, with Folan of the *Universidad Autónoma de Campeche* as the leader, has investigated different topics related to Calakmul: architecture, ceramics, climate, and lithics. They excavated some structures of the Great Plaza including buildings I, II, III and VII.

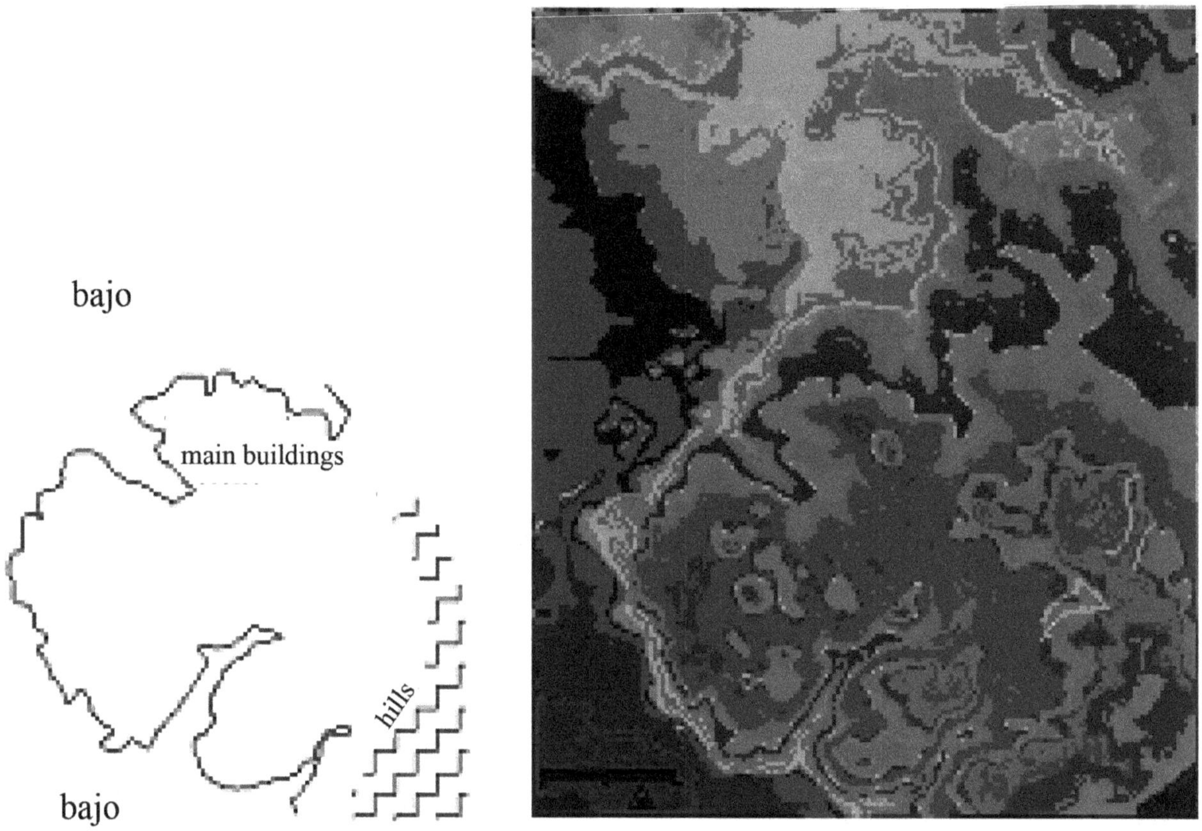

Figure 2.5. Digital elevation model of Calakmul tri-lobed promontory. The left line delineates the promontory tri-lobed form. The scale represents 1 km.

During the last fifteen years, the archaeologist Ramón Carrasco from the Mexican *Instituto Nacional de Antropología e Historia* has developed the Calakmul Archaeological Project with its main focus in the civic and ceremonial center buildings, the excavation of elite residential complexes, the study of the magnificent tombs they found, the decipherment of hieroglyphic inscriptions as well as the study and conservation of mural painting. The main goals of this project have been the restoration of some buildings within the monumental area and a detailed assessment of architecture, stelae, ceramics, mural painting and urbanism (Carrasco et al., 1999a). In 1993, with the support of the Campeche state government, he initiated the Calakmul Biosphere Regional Archaeological Project that included the study of the sites of Calakmul, Balamku and Nadzcaan.

The structures that have been continually working are I, II, IV, V, VI, VII, VIII, XI, XIII, XIV, XV, XVII, XIX, XX as well as the complexes *Chan Chi'ich*, *Kinich Pak*, *Wau Ahau Nah*, *Utsiaal Caan*, *Taman*, Northeast Group and Northern Acropolis. They have also found impressive tombs as 1, 2c, 3, 4, 6 in Sub-structure IIB or Tomb 1 of Structure XV.

- Phases of Calakmul

As at other Mesoamerican sites, the studies of ceramics have been useful to associate archaeological vestiges with particular time spans. The ceramics of Calakmul have been studied extensively. Literally tons of sherds (*tepalcates* in Mexican Spanish) have been carefully analyzed. Archaeologist María del Rosario Domínguez proposed the first classification scheme and types for Calakmul (Domínguez Carrasco, 1994). Later, Sylviane Boucher and a team have worked with the ceramics for several years identifying many new types. Calakmul ceramic complexes are listed in Table 2.1.

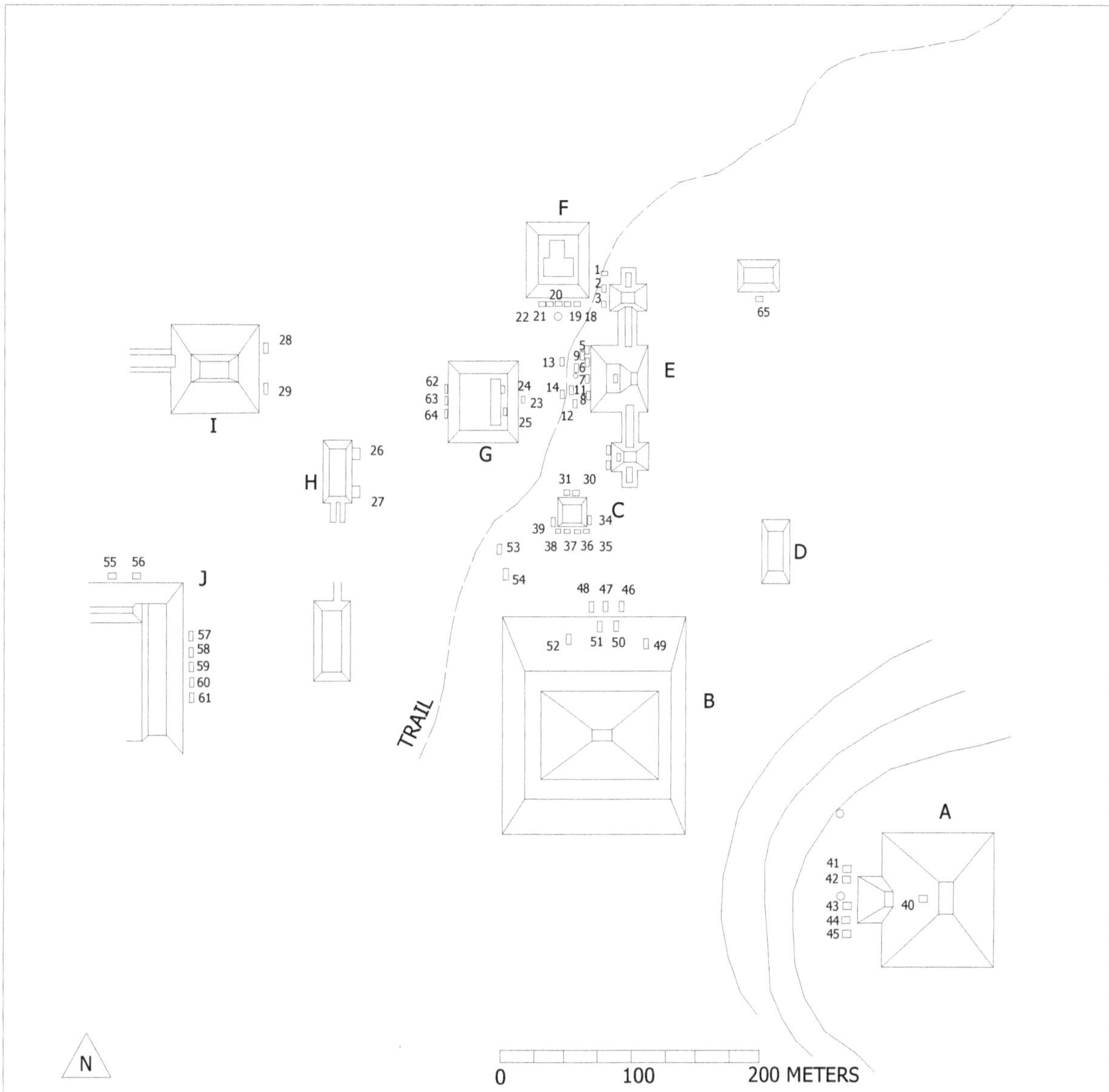

Figure 2.6. First map of Calakmul civic-ceremonial center (Redrawn from Lundell, 1933:152; Figure 2).

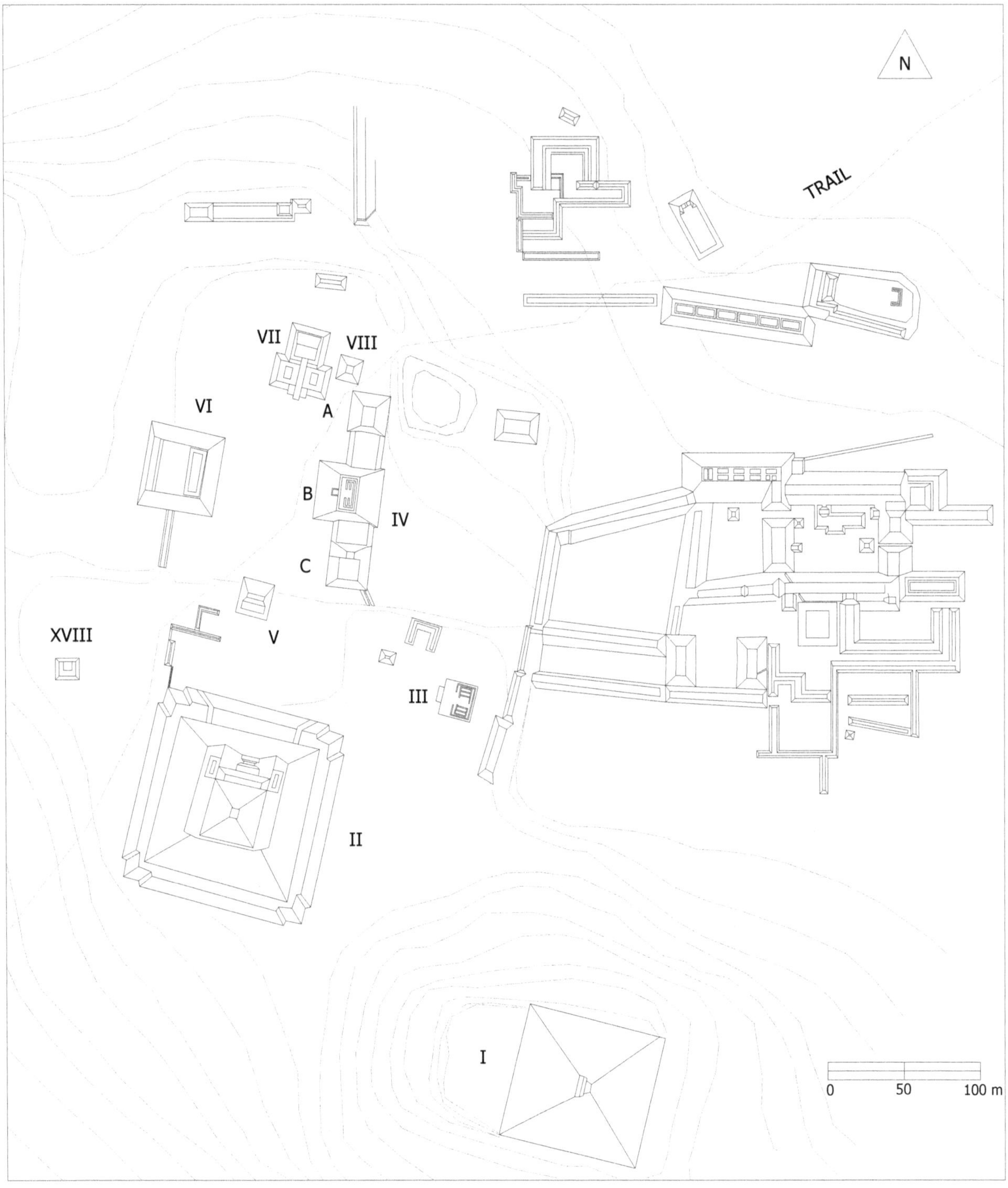

Figure 2.7. O'Neill's map of Calakmul civic-ceremonial center (Redrawn from Ruppert and Dennison, 1943).

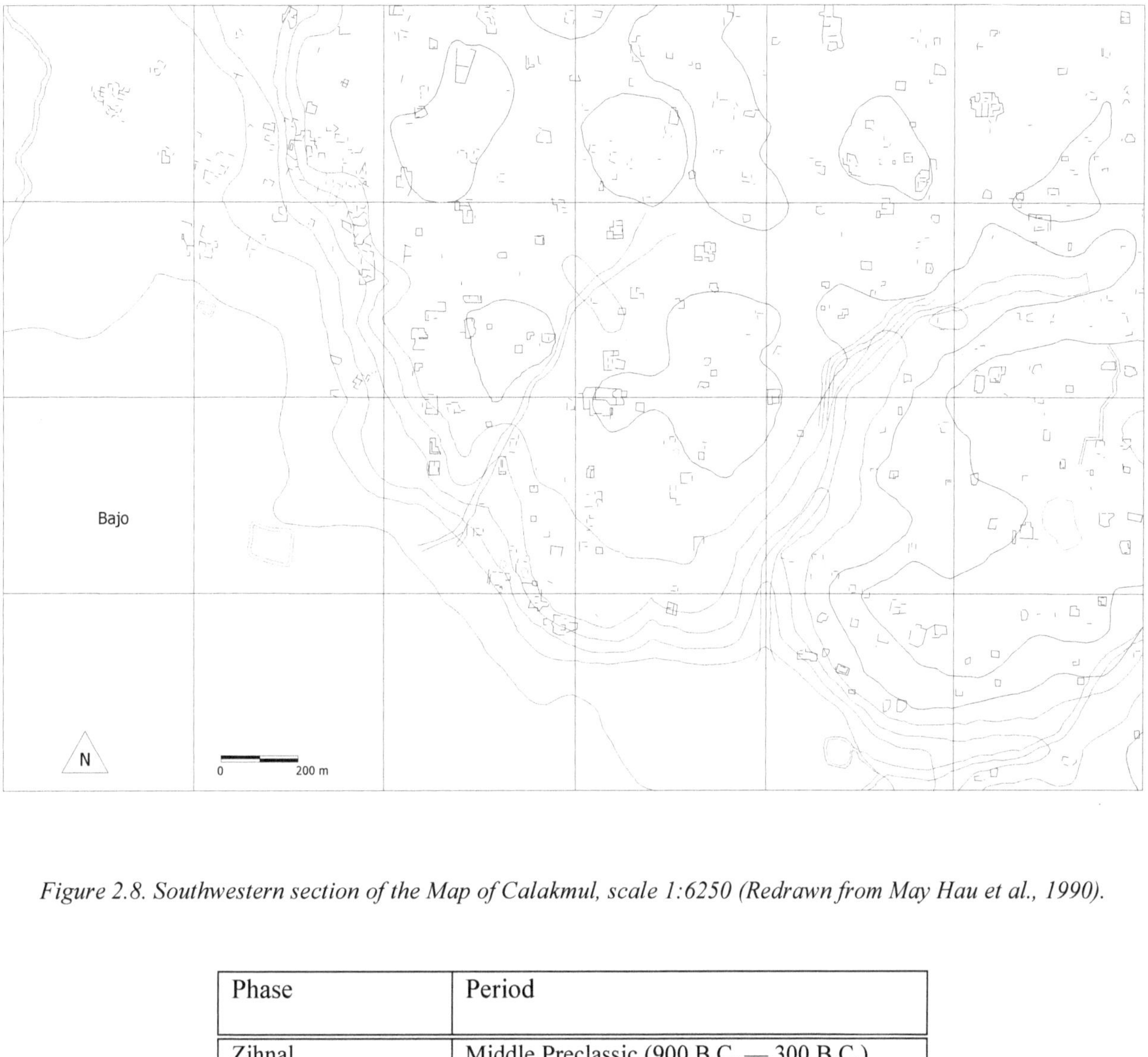

Figure 2.8. Southwestern section of the Map of Calakmul, scale 1:6250 (Redrawn from May Hau et al., 1990).

Phase	Period
Zihnal	Middle Preclassic (900 B.C. — 300 B.C.)
Takan	Late Preclassic (300 B.C. — 150 A.D.)
Not established	Protoclassic (150 A.D. — 250 A.D.)
Kaynikte	Early Classic (250 A.D. — 600 A.D.)
Ku	Late Classic (600 A.D. — 800 A.D.)
Halibe	Terminal Classic (800 A.D. — 1000 A.D.)
Cehache	(959 A.D. — ?)
Not established	Late Postclassic (1200 A.D. — 1450 A.D.)
Not established	(1450— 1800 A.D.)

Table 2.1. Ceramic complexes of Calakmul (Carrasco et al., 1996; Domínguez Carrasco, 1996).

The ceramics of Calakmul reflect the influence of its neighbors (Carrasco et al., 1996). Throughout the occupation of the site, the influence of Peten style is observed in the material. In addition, during the Classic period an affiliation is exposed with sites located in the Pasión River region, Belize, Rio Bec and Chennes area. During the end of the Classic period, sites from north and northeastern Yucatan and southern Campeche influenced Calakmul ceramics. Finally, during the Late Postclassic period only the sherds of ritual offerings and anthropomorphic censers are found.

The majority of ceramics of Calakmul were elaborated locally. Imported vessels are rare and the chemical profile of the ceramics is consistent. This indicates a strong local tradition with specialized workshops for the elite of Calakmul (Carrasco et al., 1998). This information and the fact that codex style vessels sherds have been found in Greater Acropolis buildings refute the hypothesis that codex-style ceramics were only produced in Nakbe, Guatemala (Carrasco and Colón, 2005).

In the same way, ceramic figurines were manufactured locally. The results of an archaeometric study of representative Calakmul Late Classic figurines showed that the majority was elaborated at the site with a calcium carbonate clay matrix. In contrast, the provenance of a small group of figurines made up of dolomite clay has not been identified (García-Heras et al., 2006).

- Calakmul Inhabitants

Little is known about the original settlers of Calakmul. The earliest recovered materials belong to Middle Preclassic Period when Calakmul already seems to have been a complex site. Probably the first settlers were small groups of people who lived in small villages attracted by places with water and other subsistence resources, such as the wetlands. As time went by, this society grew in complexity as happened with other Mesoamerican sites.

During the Middle Preclassic Period sites like Nakbe, El Mirador and Uaxactun, to the south of Calakmul, had also a flourishing culture. The inhabitants of Calakmul designed the first planning of the site with impressive buildings such as the early Structure II Sub C. This structure, 12 meters high dates to between 390 and 250 B.C. It is composed of three buildings, similar but not the same as to Uaxactun Group H (Rodríguez, 2003). Although there is solid evidence of an occupation during the Middle Preclassic, not much is known about it. The sites surrounding Calakmul (for example Sites 2 and 4 of the Archaeological Road Rescue Project) also have material from this period (Morales Lopez, 1985).

The construction of structures increased during the Classic Period, when Calakmul was the major rival of Tikal. In this period the rulers of Calakmul established alliances with sovereigns of the Usumacinta basin and Petexbatun region forming a broad alliance network (Carrasco, 2000). After this, the city was abandoned around A.D. 909 (Martin, 2000). During the Postclassic, the civic-ceremonial center was visited for ritual purposes, as can be inferred by the anthropomorphic censers left in the buildings (Boucher and Palomo, 2000).

- Calakmul Rulers

Maya glyphs inscribed on structures, stelae, vases, and diverse objects have been a turning point for interpreting the history of Calakmul. Researchers have found that the modification of buildings coincided with the commemoration of important events, so specific structures and tombs are related with the rulers who built or modified them.

In the same way, stelae are the vivid expressions of the most important Maya historical events. But as anyone who visits the site can appreciate, the majority of Calakmul monuments are eroded. This fact greatly limits the historical data that can be gleaned from Calakmul. However, epigraphers have been able to read some of stelae and to interpret a partial history of Calakmul from the texts that are legible.

It has been a challenge for epigraphers to track down the first rulers of Calakmul. A list of kings is represented in eleven codex style stunning dynastic vases. The vases represent nineteen rulers who presided over Calakmul, being the first one Sky Breeder (Martin and Grube, 2002:102). However, the sequence of kings is different from vase to vase, and the rulers and related events do not match with the ones represented in Maya monuments. In addition, their Calendar Round dates does not always correspond with the Maya Long Count, and cannot all be determined (Martin and Grube, 2002:102). For these reasons Martin, has proposed that this list may have a mythical context (Martin, 1997).

Although Calakmul has 120 stelae (up to the present, contains more stelae than any other Maya site), the majority is in bad condition because of limestone friability and looting. As a consequence, the information is fragmentary. In the case of some kings, names can be read, in others, epigraphers have given arbitrary names, as for example, Ruler Y, or a partial alias, *Yuknoom* Head.

Martin and Grube have reported 19 rulers, the first and second without a defined date, probably form the Preclassic period (Martin and Grube, 2002). Calakmul rulers governed from before the sixth to the tenth centuries (See Table 2.2). If we observe the list of kings, it is interesting to notice that Calakmul rulers are mentioned in sites as far away as Quirigua, Dzibanche, Seibal or Piedras Negras. It is impressive how closed the links were with distant cities.

As can be seen in Table 2.2, only five governors of the nineteen identified rulers of Calakmul have clearly defined periods of rulership. Three rulers governed for a significant number of years. Three other rulers governed for only short periods. *Yuknoom, Ch'een II* and *Yuknoom Yich'aak K'ahk* are the most mentioned kings.

The most significant events between lowland Maya kingdoms recorded in inscriptions are (Martin and Grube, 1995):

A.D. 514: Calakmul was overlord of Piedras Negras.
A.D. 545: A ruler of Naranjo acceded to the throne "by the doing of" a Calakmul Ruler.
A.D. 631: Calakmul joined Caracol to defeat Naranjo.
A.D. 648: Calakmul was overlord of Ruler 1 from Dos Pilas.
A.D. 599: Calakmul attacked Palenque.
A.D. 611: Calakmul attacked Palenque.
A.D. 656: Calakmul is involved in the accession of a ruler from Cancuen.
A.D. 657: Calakmul attacked Tikal.
A.D. 677: Calakmul is involved in the accession of a ruler from Cancuen.
A.D. 693: Calakmul was overlord of Ruler Smoking Squirrel from Naranjo.
A.D. 695: Calakmul was defeated by Tikal.
A.D. 743: Tikal defeated El Peru, Calakmul vessel.
A.D. 744: Tikal defeated Naranjo, Calakmul vessel.
A.D. 909: *Aj'Took'*, new *ajaw* Kan.

The Early Classic was a period in which few events were registered. The Stela 114, the earliest one of the site (A.D. 435) is erected (Pincemin et al., 1998), as well as Stela 43 (A.D. 514). Kings as *K'a'tuun Hix*, Sky Witness, First Axewielder and Scroll Snake governed Calakmul. In this time there was a patent control over Piedras Negras and Naranjo. It has been proposed that the lack of inscriptions during this period for 109 years suggests the destruction or the movement of the Snake capital seat to another site (Martin and Grube, 2002).

The Late Classic Period was an intense time for Calakmul. It was the epoch of the great rulers: *Yuknoom Ch'een I*, *Tajoom Uk'ab'K'ahk*, *Yuknoom* Head, *Yuknoom Ch'een II*, *Yuknoom Yich'aak K'ahk'*, Split Earth, Rulers Y and Z as well as *Bolon K'awiil*. The personage buried in Tomb 4 of Structure II of Calakmul was probably *Yuknoom Yich'aak K'ahk'*, as can be read in a polychrome plate found next to him. Tomb 4 is almost certainly the tomb of this king, despite the fact that a Tikal inscription (Lintel 3, Temple 1) states that he was captured and sacrificed in that place in 695 A.D. (Carrasco et al. 1999b).

During Late Classic, Naranjo was conquered and there was a war with Palenque. Calakmul and Caracol were allies. Flannery (Flannery, 1972), Marcus (Marcus, 1973) and Adams (Adams, 1986) have considered Calakmul as a regional capital of the Late Classic, as important as Tikal, Palenque and Copan. Marcus proposes Calakmul as a center that controlled Naachtun, Oxpemul and La Muñeca (Marcus, 1973). Inscriptions found all over Maya Lowlands point out to two so called "superstates" that were Tikal and Calakmul (Martin and Grube, 1995). During this period it is clear that Calakmul served as the seat of the Snake polity (Martin, 2005).

During the Terminal Classic the splendor of the reign was decaying. The rulers *Chan Pet* and *Aj Took'* were the last ones from a splendid dynasty. Finally, during the Postclassic period, there is evidence of a group called the Cehaches who reportedly inhabited a large area from Chuntiuqui Guatemala to Bolonchen, Campeche until the seventeenth century. These people might have been the last descendants of the inhabitants of Calakmul (Carrasco, 2000:13).

- Description of the Site

Calakmul is located in a tri-lobed promontory that measures about 17 km^2. The *Bajo* El Laberinto surrounds almost all this prominence. To the north, east and west a principal seasonal watercourse (*coriental*) follows the limits between the *bajo* and the promontory. The site center with the principal temples and palaces measures about 1.4 km^2 comprising an elliptical form.

The archaeological site has 6345 stone and wood thatched structures distributed in a disperse pattern (Folan et al., 2001a:48). They have been classified into various types: ceremonial or civic complexes, elite residences limited by walls, and small complexes (probably the commoners' dwellings for living). The architectonic style of Calakmul corresponds to Peten style, with vaulted temples decorated with polychrome stucco over big massive platforms. Calakmul is the Maya city with more stelae inscribed (120) than in any other site. These stelae date to between 435 and 909 A.D.

The extent of Calakmul kingdom has been the subject of some debate. Adams proposed that as a state, it comprised an area of 15 000 km^2 (Adams, 1986). Folan proposes that this regional state had more than 13 000 km^2 (Folan et al., 2003) with a population of 1 500 000 inhabitants (Braswell et al., 2004). Other estimates suggest that the number of residents of the city was 20 000 (Fletcher and Gann, 1987).

Maya Name of the Ruler	English Name of the Ruler	Period of Government	Monuments and artifacts with rulers mentions
?	Sky Breeder	?	Dynastic vases
Yuknoom Ch'een I	?	?	Dzibanche
K'a'tuun Hix	?	>520 - 546	Naranjo
(u) - ? - [?] - *Chan-na*	Sky Witness	>561 - 572	Los Alacranes, Caracol, Yo'okop
Yax ? – Yopaat	First Axewielder	>573	Dzibanche
Uneh Chan?	Scroll Snake	579 - 611	Caracol
Yuknoom Ti'Chan	? Sky	>619>	Caracol
Tajoom Uk'ab'K'ahk	Burning Hand	622 - 630	Caracol, Naranjo
Yuknoom II	*Yuknoom* Head	630 - 636	Naranjo
Yuknoom Ch'een II	*Yuknoom* the Great	636 - 686	Calakmul, Dos Pilas, El Perú, Cancuen, Moral, Piedras Negras, La Corona
Yuknoom Yich'aak K'ahk	Fire Paw	686 - 695?	Calakmul, El Perú, Dos Pilas, Tikal
?	Split Earth	695?>	Tikal
Yuknoom Took'K'awiil	Flint God K	>706 - 731	Calakmul, Dos Pilas
Wamaw K'awiil	?	>736>	Quirigua
?	Ruler Y	>741 >	Calakmul
?	Ruler Z	>751>	Calakmul
Bolon K'awil	Nine God K	>771 - 789?>	Calakmul, Tonina
Chan Pet	?	>849>	Seibal
Aj Took'	He of Flint	>909>	Calakmul

Table 2.2. The nineteen rulers of Calakmul. Table with their names, government period, monuments and artifacts with ruler mentions (Martin, 2000; 2005; Martin and Grube, 2002).

Calakmul has been considered as a first-order site when compared with others, because of its extension, monumental buildings, elevated number of courtyards between other characteristics (Adams and Jones, 1981). The city is arranged in disperse architectonical complexes that surround the Great Plaza. There does not seem to be a uniform pattern of urban distribution. At least eight large complexes can be appreciated in the map of Calakmul. It seems that constructions were adapted to the topography.

Ruppert divided the city into various spaces and groups: the Central Plaza, West Group, East Group, Southeast Group and Northeast Group (Ruppert, 1943). At present, the archaeologists of *Proyecto Arqueológico Calakmul* divided the civic-ceremonial center into the Great Plaza, Great Acropolis, Northern Acropolis, Small Acropolis, Southeast Group and West Group. These last five architectural complexes surround the Great Plaza. The most intensively studied part of the site has been the Great Plaza comprising structures II to VIII. This group has one of the highest buildings of the site, and some of them have tombs. During the recent years excavations have been conducted also in the other Acropolis.

Structure II is a good example of how Calakmul went through great changes in the course of its history. Structure II is the only building that has an architectonic sequence from the Middle Preclassic to the Terminal Classic with an occasional occupation in the Early Postclassic (Carrasco, 2003). At present, its 12-meter Middle Preclassic substructure, II Sub C, constitutes the most ancient rubblework building in the Peten area. This barrel or tunnel vaulted structure has a front façade with impressive iconography. A celestial band frames a scene of two personages with zoomorphic characteristics that surround another personage that is descending. During the Late Preclassic another building of 48 meters high was constructed above, probably with triadic temple arrangement on top (Rodríguez, 2003). In Early Classic times, two more superimpositions were added: structures IIC and IID in the top with their respective perrons and big masks with a different stylistic pattern (Rodríguez, 2000). During this period the stela 114 was erected (A.D. 435). In the Late Classic, a final modification was made: masks were covered and Structure II B was erected with its perron (Rodriguez, 2000; 2003).

In Structure IIB, the Tomb 4 of the ruler *Yuknom Yich'aak K'ahk* (Fire Claw) was found. The body was wrapped with successive layers of palm with fine clothes placed on a wooden bier. Several jadeite adornments, a marvelous jadeite mask, shells, and ceramics were some objects that were buried with the lord. As companions, two individuals were placed next to the principal chamber. In Tomb 6, a women and a young male were buried with ceramics utensils, adornments and ritual objects (García Moreno and Granados, 2000).

In addition to architecture, impressive vestiges have been located in Calakmul as astonishing jade masks, beautiful ceramics, shell implements, lithics, and in the last years, magnificent mural paintings (Carrasco and Bojalil, 2005). All these remains attest to the greatness of the long-lived Calakmul kingdom.

It has been proposed that many roads, called in Maya *sacbeob*, communicated Calakmul with its neighbors. Folan and colleagues (2001b) identified fifteen of them in the site and surroundings. There were regional, state, urban and neighborhood roads according to their length and connection. Given the rainforest surroundings, *sacbeob* are difficult to identify. Personally, I was not able to find any of the *sacbeob* that May Hau, Canche and Chan report (Folan et. al., 2001a) in the areas we walked. In addition, archaeologist Ramón Carrasco and the members of the Calakmul Archaeological Project have not seen them (Carrasco 2003, personal communication). Since these roads are sometimes difficult to identify at ground level, I also tried to find them in the aerial photographs, but without success.

However, through aerial photographs I was able to detect a feature that resembles a road that was not identified by Folan and his colleagues (Folan et al., 2001a). This feature with a length of 5.18 km goes from northeast to southwest inside the El Laberinto *bajo* reaching the promontory near the west section of the civic-ceremonial center (Figure 2.9). Actually, it points directly to the site center. It is plausible that this linear feature is a road but it could be modern or Pre-Columbian; it still needs to be checked on the ground.

It seems interesting that a road could cross El Laberinto. And if it were indeed a path then it would be indirect evidence that this *bajo* was human-modified. Since the *akalche* gets inundated during the rainy season, a road there could be evidence of a seasonal pass or that water was directed to planned places such as reservoirs or channels. This subject will be discussed in the other chapters of this study.

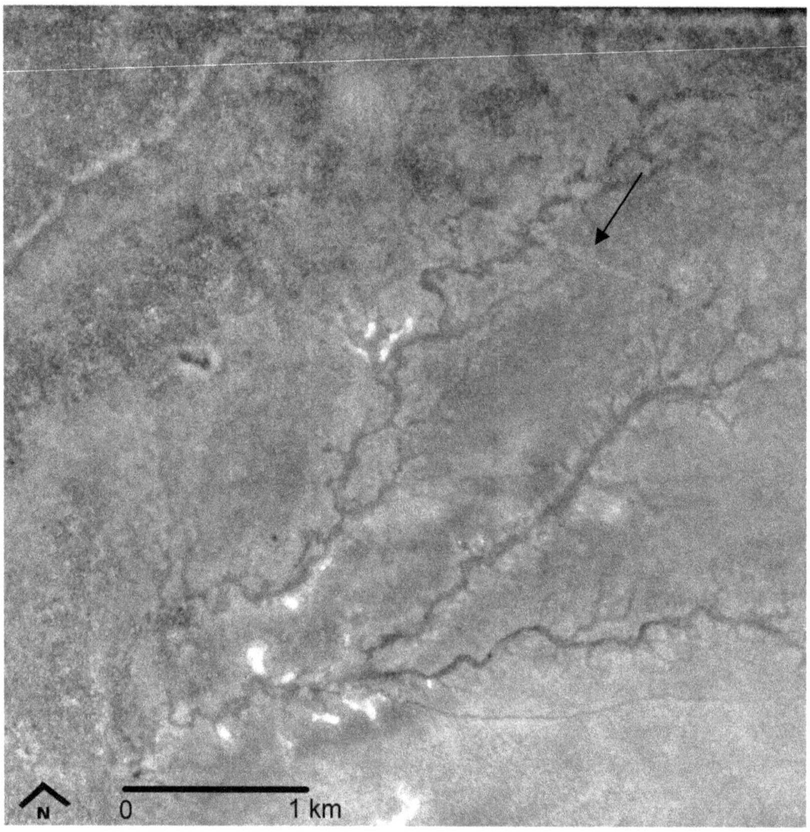

Figure 2.9. Aerial photograph showing a possible road in El Laberinto Bajo (INEGI, 2001a). The arrow points to the linear feature that has northwest-southeast direction.

The Extension of Calakmul and its Catchment Area

As the site was explored, ruins throughout Calakmul were observed except in the area of the *Bajo* El Laberinto. This fact is evident in the map of Calakmul (May Hau et al., 1990; see Figure 2.8). The densest area of construction is located to the north of the promontory. Ruins can also be observed along the edges of El Laberinto, as Folan has pointed out (Folan et al., 2001a). Our group did not detect any structure in El Laberinto, but Folan reports that some buildings, possibly used in the processing of flint, were scattered in the *akalche*. He also affirms that his team excavated in the *bajo*, and they did not encounter buried ruins (Folan et al., 2001a).

Some natural boundaries are evident in Calakmul. Although they are not impassable, these limits are well defined with respect to the core area. Hence, I propose them as the boundaries of the site and its periphery (Figure 2.10). The first boundary that defines a particular space is the *corriental* (large intermittent stream) that encloses the promontory to the north, northeast and west. This stream disappears in the dry season, but in the rainy season it channels enormous quantities of water. In fact it was possible to observe remains of tree trunks, dead leaves and rubbish in tree branches at a height of over 2 meters. At some points, the *corriental* becomes flat and wide, in others, deep.
The limits posed by the *corriental* are reinforced by the Bajo El Laberinto, which surrounds the promontory to the north, east and west. During the rainy season, a significant area becomes inundated, enhancing the limits. To the south, the gradually elevated hills of the karstic spine form another limit.

In sum, it is suggested that these natural features, that enclose an area of 17.27 km^2, delimit specifically the city and its periphery (*Uxte' tuun*). This proposal is also supported by the concept of catchment area. Vita Finzi and Higgs have suggested that the prime land of a site is located within a radius of 5 kilometers from its ceremonial center (Vita Finzi and Higgs, 1970). In their site catchment area study, they propose that every settlement has a resources zone for satisfying primary needs that occur within a reasonable walking distance of 5 km. They also state that it is inefficient for people to walk more those 5 kilometers to satisfy their basic needs.

In order to graphically exemplify the catchment area of Calakmul, a series of concentric circles one kilometer apart are shown in Figure 2.10. Structure II of the civic-ceremonial center is considered to be the center of the circles. The most important buildings are located within the first circle (site center) with an area of 3.14 km^2, and the promontory itself falls within the next two (called neighbourhood and periphery). The neighbourhood area mesures 9.42 km^2 and the periphery 15.71 km^2. The majority of habitational units are found in them. In sum, the areas exhibiting dense construction are within the area delimited by the first three circles (28.27 km^2).

Further afield, in proximate rural and rural area, with 21.99 and 28.27 km^2, respectively, only scattered platforms are encountered. In this work, the first four circles (radius of 4 km from Structure II; 50.26 km^2) will be considered for further analyses. Since it is convenient to work with a square frame, the study area will comprise 64.6 km^2.

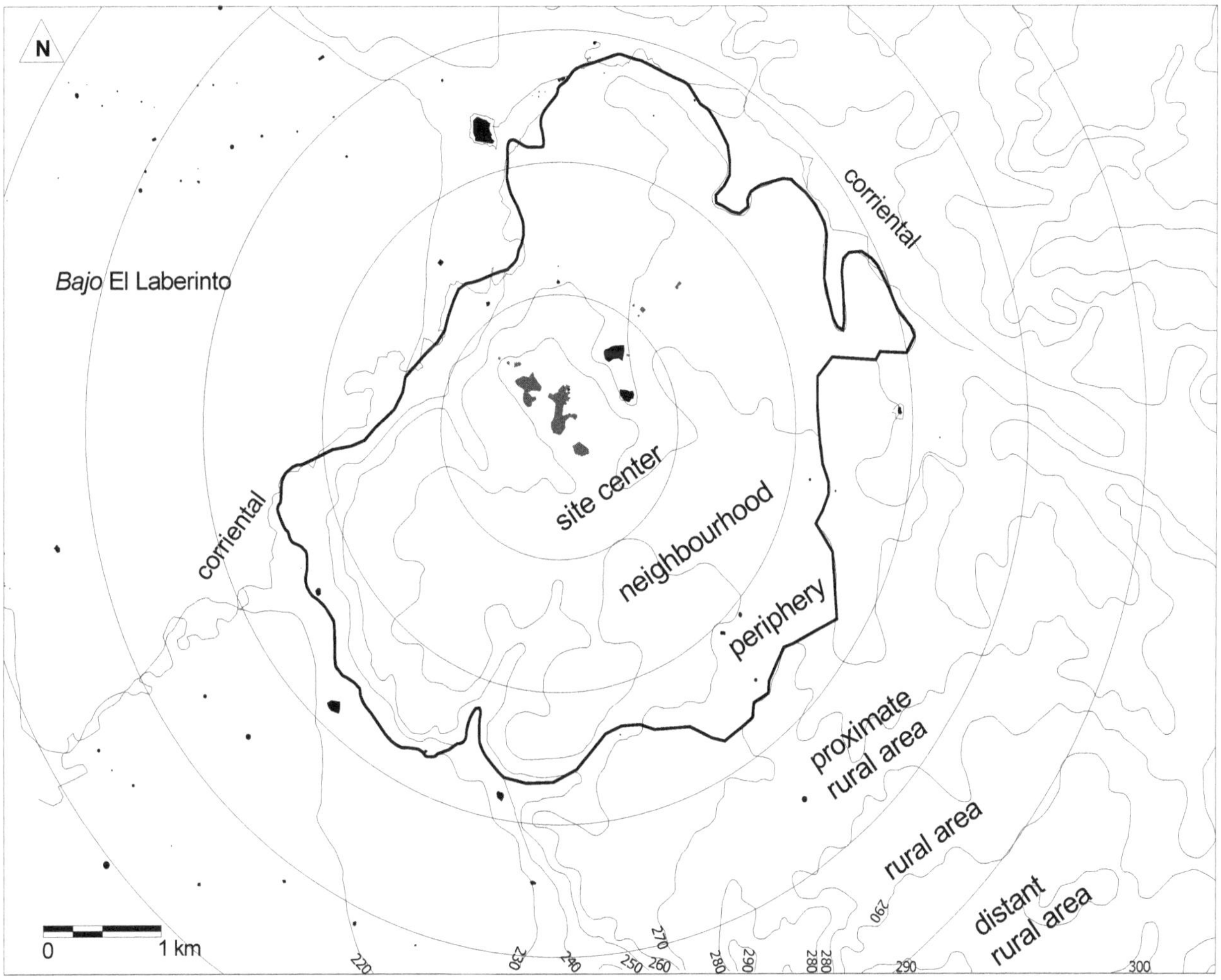

Figure 2.10. Boundaries and site catchment area of Calakmul. Each circle has an increasing radius of 1 kilometer. Major buildings are shown in light gray, reservoirs in black and the proposed boundaries of the city and its periphery with black line.

CHAPTER 3. THE PHYSICAL SETTING OF CALAKMUL

"Yucatan is the country with least earth that I have seen, since all of it is one living rock and has wonderfully little earth, so that there are few places where one can dig down without striking great layers of very large rocks. The stone is not very good for delicate carvings, as it is hard and coarse, but such as it is, it has been sufficient for their having made of it the great number of buildings, which there are in that land. It is very good for lime of which there is a great deal, and it is marvelous that the fertility of this land is so great on top of and between the stones."

Fray Diego de Landa in Tozzer, 1941:186.

Location of Calakmul

The ancient city of Calakmul is located in the municipality of Calakmul, in the southern Mexican state of Campeche, about 30 km north from the Guatemala frontier (Figure 3.1). The coordinates of the site are UTM 16Q 202560 2003883. It is possible to reach the site following the federal highway No. 186, turning to the south at Km. 96 and driving 68 km on a paved road through the jungle.

The archaeological site is within the Calakmul Biosphere Reserve with 723,185.12 hectares in extent, which comprises several additional archaeological sites and impressive natural areas (Folan y García Ortega, 2000; Figure 3.1). The reserve, decreed in 1989, has two nuclear areas and a buffer surface that surrounds them (García-Gil et al., 2002). The purpose of the nuclear areas is to protect the ecosystem from any human impact. The extension of the northern one measures 100 345 ha and the southern one, 147 915.5 hectares (SEMARNAP, 1995). The aim of the buffer area is to restrict human subsistence activities in order to maintain the nuclear surface as untouched as possible. This transitional area measures 474 924.62 ha. In 1993 this reserve was included as part of the UNESCO Man and Biosphere program (SEMARNAP, 1995). Overall, protected areas in southern Yucatan Peninsula will be enlarged with the creation of the reserve of Balamkin (10 000 ha), Balamku (409 000 ha) and Balam Kaax. All the mentioned reserves, in conjunction with protected areas in Guatemala and Belize, will sum about 2 000 000 hectares (García-Gil et al., 2002).

People from elsewhere in Mexico have colonized the biosphere margins over the past thirty years. (Escamilla et al., 2000). During the 1970s, as part of the Agrarian Reform, President Luis Echeverría handed over land in Tabasco, Chiapas, Quintana Roo and Campeche. As a consequence, people from all the country, many of them without agricultural knowledge, colonized the land (Gurri, 2006). At present, the estimated population of *Municipio* of Calakmul is 30 329 inhabitants (CONAPO, 2006). Although much forest area has been protected, researchers expect that soil use will inevitably change because of the expansion of the agricultural and livestock frontier (García-Gil and Pat, 2000).

Physical Characteristics

The site of Calakmul is positioned on a semi-rounded prominence that extends beyond a series of hills with a northwest-southeast direction. The site is located on a karstic concave surface with steep ravines toward the west and south that divide the promontory in three parts. El Laberinto *bajo*, a seasonally flooded plain area, surrounds almost three quarters of the prominence (Figure 3.2). The archaeological site's major buildings were erected taking advantage of the hills of the prominence. In contrast, the multiple small platforms, presumably the household areas, were built in concave and flat terrain.

- Climate

The climate of Calakmul is classified as Ax' (w_1) (i_1) gw'' (Orellana et al., 2003). The environment is warm, tropical sub-humid, with copious rains between June and November, with a dry month between July and August called in Spanish *canícula*. The temperature, humidity and pluvial registers from the nearby meteorological station are shown in Figures 3.3 and 3.4. The location of the Calakmul station is 18° 21' 54" N and 89° 53' 33" W, with an altitude of 25 m. This station has automatic records each 10 minutes since 2003 (Comisión Nacional del Agua, 2007).

The annual average temperature is 25.2 °C (Comisión Nacional del Agua, 2007). The coolest months are November, December, January and February, followed by a dry period in April and May with temperatures up to 28 °C that decrease gradually during the following months (Figure 3.3). Annual humidity varies from 70 to 90%. The drier months are February, April and March, with values between 60 to 70%, and the most humid period occurs during winter with more than 80% of humidity (Figure 3.3).

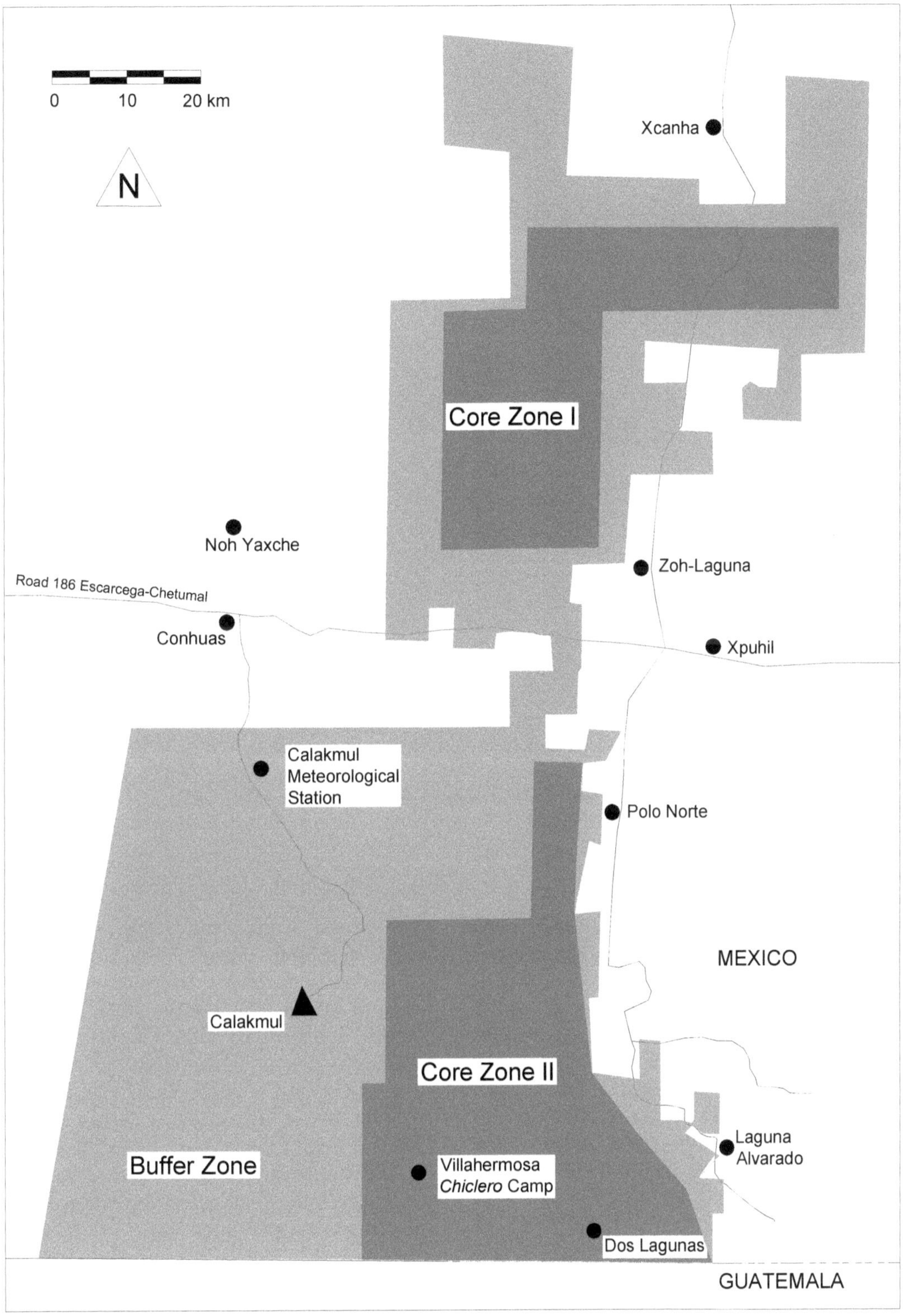

Figure 3.1. Map of Calakmul Biosphere Reserve with main towns and roads (based in García-Gil et. al., 2002:41). The dark gray colored surface represents the extension of the two biosphere nuclear areas, surrounded by the buffer area in light gray. Notice that the reserve reaches the Mexican frontier with Guatemala.

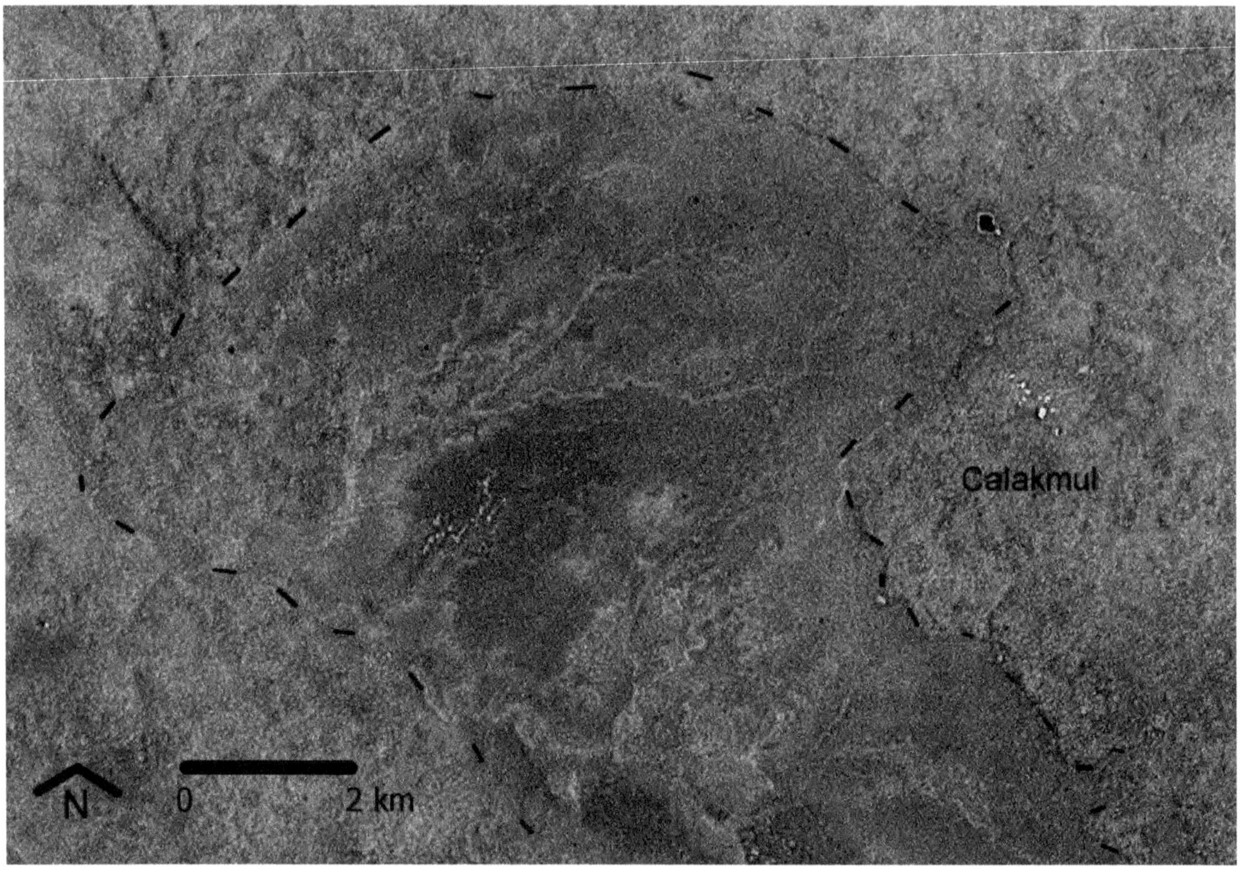

Figure 3.2. Landsat image of Bajo El Laberinto and Calakmul. Notice that the upper part of El Laberinto has a partial elliptical form. The black irregular square located to the north of Calakmul civic-ceremonial center is the big aguada.

The annual average precipitation at Calakmul is 948.9 mm (Comisión Nacional del Agua, 2007). The range of annual precipitation varies from 800 to 1100 mm (Figure 3.4) with fluctuations from year to year. As a consequence, farmers receive each year varying quantities of rainfall. Similarly, the observation of rainfall records in Noh Yache and Zoh Laguna located near Escárcega-Campeche No. 186 Highway (location on Figure 3.1; Comisión Nacional del Agua, 2005) as well as records in Peten and Yucatan, highlights the fact that average rainfall is not stable. In northern Yucatan, Page (1933, cited in Lundell, 1937:7) states that there are cycles of wet and dry seasons that vary greatly from year to year. This same situation was observed in northern Peten (Lundell, 1937).

The Calakmul Biosphere reserve presents a decreasing humidity gradient from southeast to northwest in the same way as precipitation. Factors such as summer rains, trade winds, autumn and winter hot-humid Caribbean winds, hurricanes promote a high rate of precipitation variability. In addition to this phenomenon, a decreasing tendency in rainfall can be noted elsewhere in the Yucatan Peninsula over the long term (Martínez et al., 2001).

The humidity of the rainforest derives from the Atlantic and Caribbean northeasterly winds. It varies depending on the Caribbean tropical storms that normally run toward the west. When the storms become hurricanes, the rain increases and large areas of rainforest become inundated. The winter rain is the result of the cold northern air masses (García Gil et al., 2002). The climate of the biosphere has been a decisive factor for impeding human inhabitation. The water shortage during the dry season restrains the occupation by large numbers of people. Today local inhabitants buy water to satisfy their personal needs.

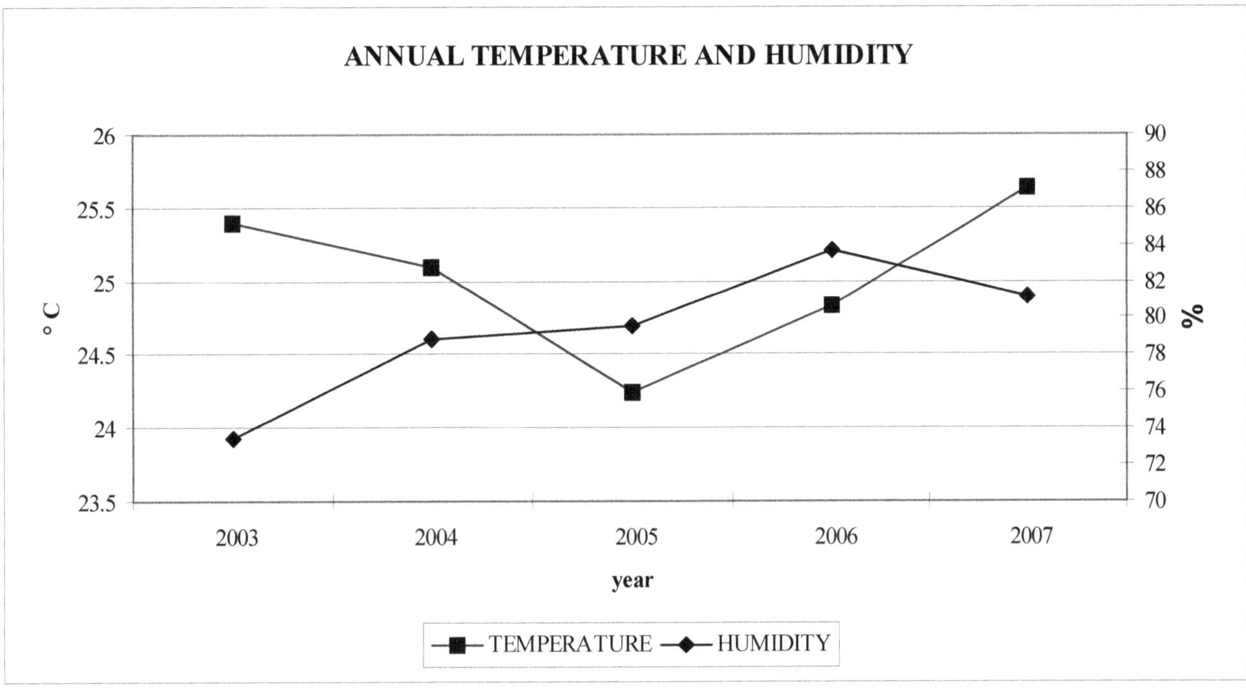

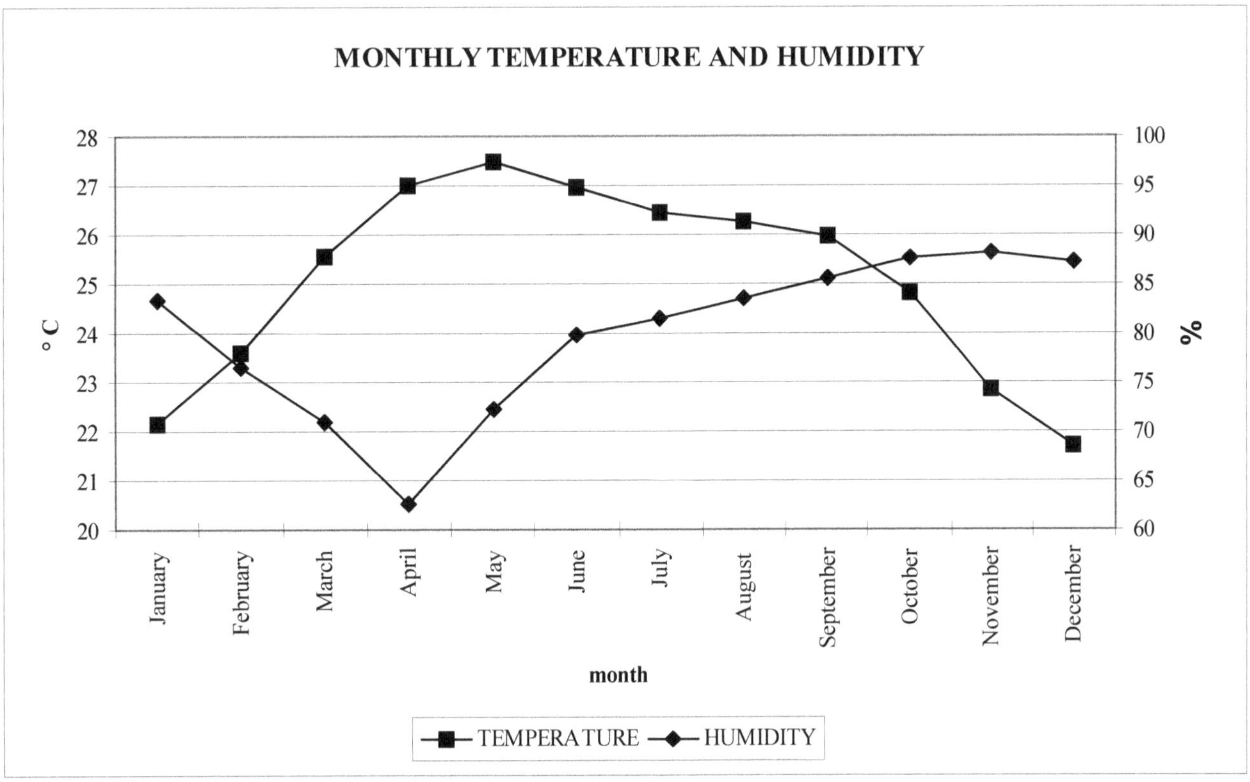

Figure 3.3. Annual and monthly temperature and humidity percentage recorded at Calakmul meteorological station (Comisión Nacional del Agua, 2005).

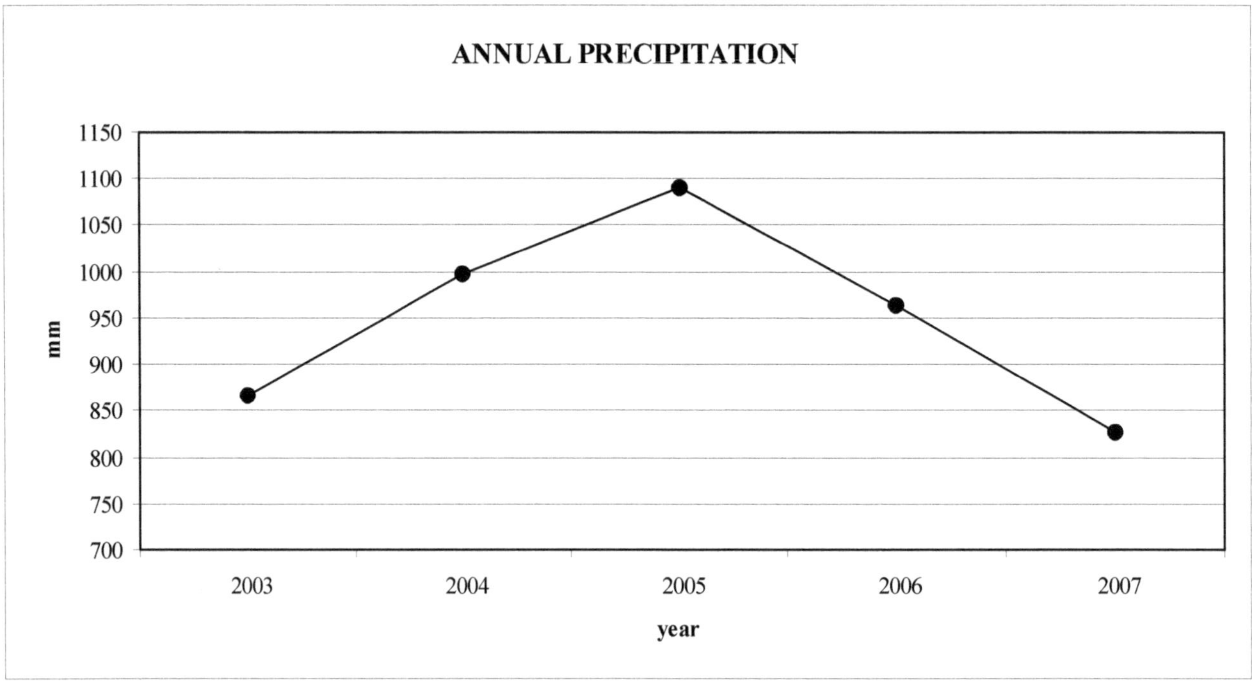

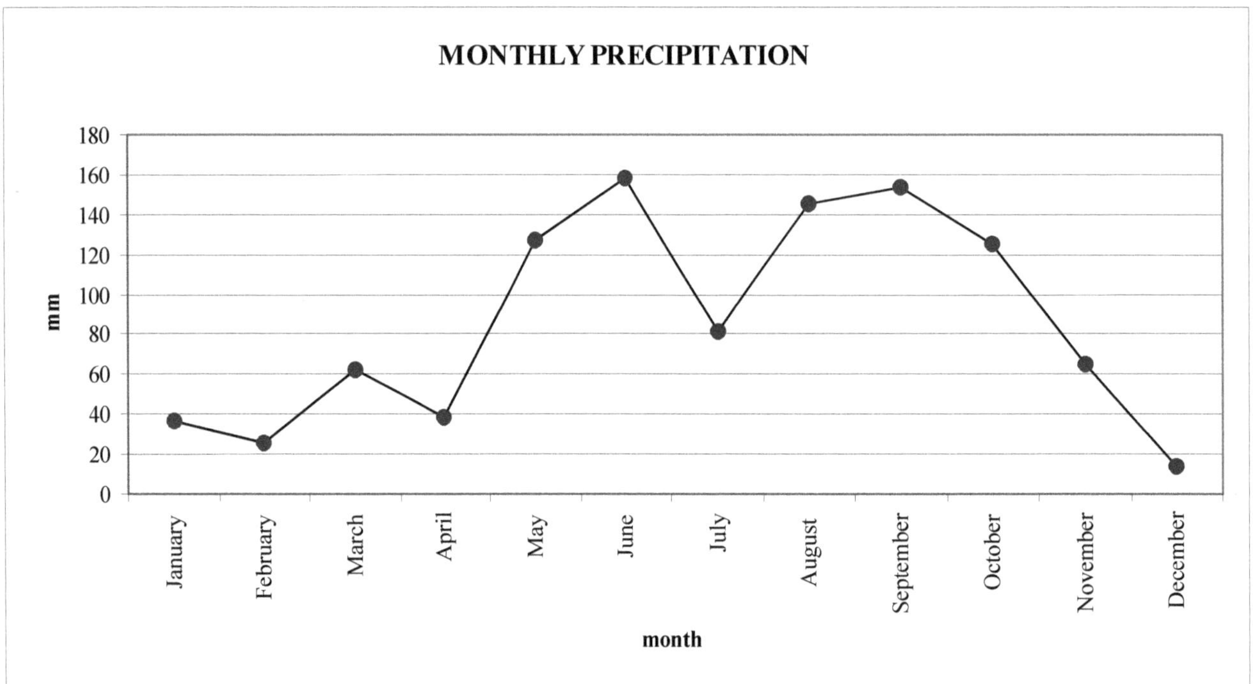

Figure 3.4. Average of total annual and monthly precipitation recorded at Calakmul meteorological station (Comisión Nacional del Agua, 2007).

- Physiography

Calakmul lies in the Yucatan Platform province and in a karst and hillock sub-province. Shallow oceans in the margins of ancient continents originally formed the limestone rocks, with several fossils (Gates, 1999). There are numerous dome hills that have suffered a continued dissolution that shaped the calcareous rocks as valleys and depressions. The main rocks that lie horizontally below a *caliche* level are limestone, gypsum, marl and dolomite (INEGI, 1999). *Caliche* is a non-geologic layer near the surface, more or less cemented by secondary carbonates of calcium or magnesium precipitated from the soil solution. It may be a soft thin soil horizon, a hard thick bed just beneath the solum, or a surface layer exposed by erosion (Research Branch, 1976).

- Geology

The parental material is from the Early and Middle Paleocene (deposits between 36.5 and 66.5 millions B.P.). These rocks are considered as microcrystalline with gypsum and anhydrite, silica or dolomite limestone, marl, as well as medium and thick layers of gypsum (Butterlin and Bonet, 1963). The anticline of Calakmul has rocks from the Inferior Paleocene and Early Eocene with carbonates, sulfates and flint beds, flanked in the east and west by recent rocks from the Eocene (Gates, 1999). The *bajo* areas have Quaternary alluvial laminar deposits with calcareous plastic clays (INEGI, 1999), as well as non-consolidated material as gravel, sand and silt (García-Gil et al., 2002).

Limestone and flint were of extreme importance because these were the main tools utilized for buildings and generally speaking, for subsistence. Calakmul had local flint outcrops in *Bajo* El Laberinto (Folan et al., 2001a). An X-ray diffraction, electronic microscopy and chemical analysis of the internal structure of the flint found in structures II and III of Calakmul revealed its homogeneity (Espinosa et al., 2001). This fact indicates that flint was extracted from the same contiguous area. A variety of implements were also made from limestone, such as *manos* and *metates* (grinding stones), construction tools, scrapers, chisels, hoes and axes (Folan et al., 2001a).

Investigators have reported up to thirty-nine quarries at Calakmul (e.g. Gallegos et al., 1985; Morley, 1933). Limestone quarries were an important resource for the people of Calakmul; raw materials extracted from them were used to build temples and houses. Gallegos excavated three quarries near the Central Plaza and two limestone outcrops (Gallegos, 1985; 1994). It seems that the Maya wore away the rock from the limestone block leaving a small piece that they probably cut off for purposes of construction, sculpture, and/or to make tools. As Ruppert comments:

"Another find, possibly more significant scientifically, was Bolles' discovery of an ancient quarry a short distance north of the Main Plaza. Here two large blocks of stone, which, judging by their size and shape, probably had been intended for use as door lintels, are only partially quarried, one end and one narrow edge of each still remaining fast to the bed of limestone from which they were being worked. While by no means unique, definitely identified quarries at Maya sites are sufficiently rare to merit comment, and the one found this season at Calakmul is perhaps as good an example of this kind yet reported from any Old Empire site" (Ruppert, 1943:23).

Clay sources were also of utmost importance. The great majority of ceramics and lithics at Calakmul were processed locally (Carrasco et al., 1998; Espinosa et al., 2001), but external influences and relations have been suggested by the manufacture of specific and distinct ceramic types. Ceramics link the site of Calakmul with the Peten during the Late Preclassic; with northeast Peten and the Río Bec region during the Early Classic; with the northeast Peten and Río Bec and Chenes regions during the Late Classic, and with the north and northeast of Yucatan, as well as the Río Bec area during the Terminal Classic (Carrasco et al., 1996).

- Soils

The soils of Calakmul belong to the Xpujil Association that groups Rendzinas, Gleysols and Vertisols (FAO, 1970). The soils are developed over calcareous rocks. The Xpujil Association sub-unit relates the soils located in Calakmul in areas between 200 and 380 meters above the sea level, with plains, concave elevations and small hills. The soils in the hills are Lithosols (*t'sekel* in Maya) and Rendzinas. Locally, Rendzinas are known as *box lu'um* if they are black with stones, *pus lu'um* if they are gray with few stones and *chi'ich lu'um* if they are brown with a granular structure (Morales and Magaña, 2001). In the depressions, Gleysols (*akalche*) and Vertisols (*ya ax hom*) can be observed.

- Hydrology

Calakmul is located in the hydrological region Number 30 Grijalva-Usumacinta, Basin C, Sub-basin g1896, that measures 1896 km^2 (INEGI, 1988) almost at the edge of the West Yucatan hydrological region Number 31, that has 25 406.13 km^2 (Díaz Ponce, 1999). There are few permanent hydrological features in Calakmul because of the high calcareous permeability and geological fractures. As a consequence, the drainage is essentially karstic subterraneous with few superficial sub-dendritic currents and high infiltration rate (INEGI, 1999).

When it rains, the majority of water evaporates and the rest infiltrates through fractures and karstic holes to the aquifer below or stays in the *bajos*. The rain permeability coefficient in the hills is 0 to 5% (high); in the *bajo*, 20 to 30% (low; INEGI, 1988). These percentages represent the water accumulation in the surface. The thick high terrain limestone cover (*caliche*) allows infiltration and absorbs rain until the saturation point, after which it spills. The flat terrain of some areas of the hills keeps relatively humid because of karstic infiltration. The huge seasonally inundated *Bajo* El Laberinto keeps water from rain for months after the rainy season finishes. This water accumulates from the rainfall and from non-permanent currents that channel water from the highest points.

The superficial hydraulic features are intermittent large and small streams, *bajos*, and *aguadas*. The River Calakmul is the closest non-permanent stream, running with a northwest direction from the hills to the depression, and parallel to the site but not through it.

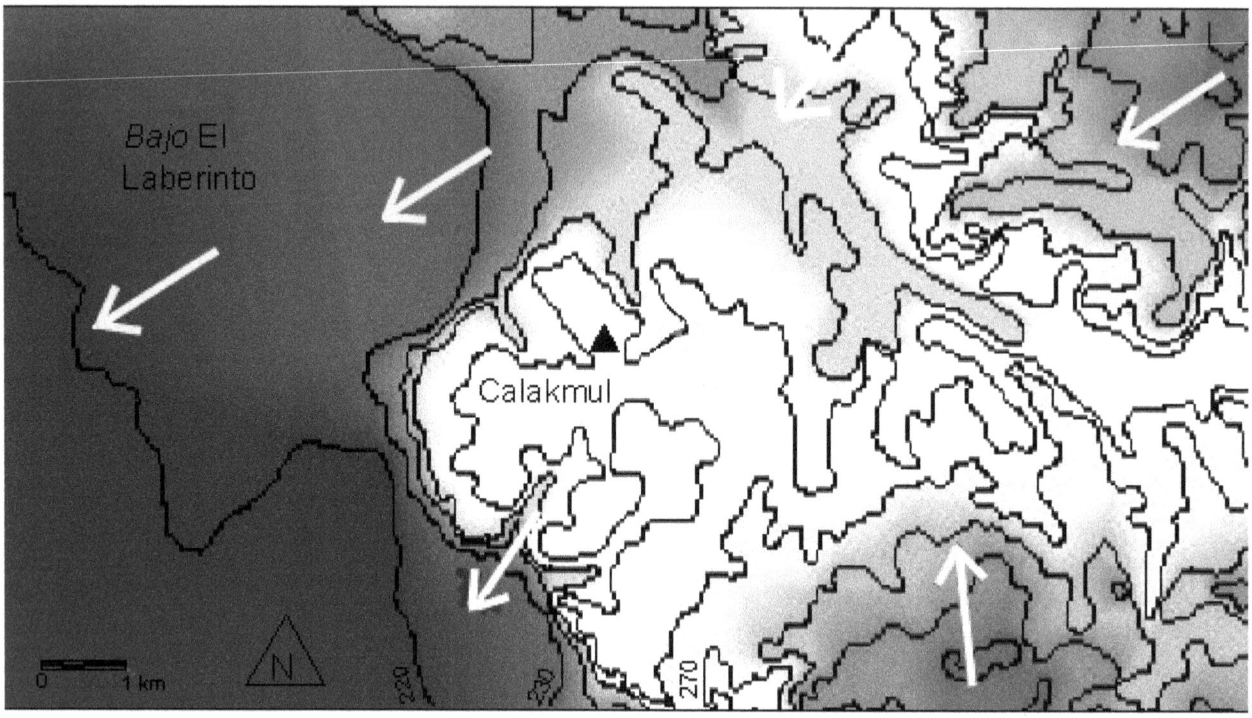

Figure 3.5. Digital elevation model of El Laberinto and Calakmul region. The arrows point to the direction of rainwater flux. The distance of each contour line is 10 m. The light gray tones show the highest areas in contrast with the dark gray tones that show lower ones. The archaeological site of Calakmul is showed by the black triangle. The scale indicates 1 km.

Particularly in the Calakmul region, the eastern adjacent hills split, and conduct all the water toward the *Bajo* El Laberinto (See Figure 3.5). Before the water reaches the *bajo*, it goes through Calakmul's site margins, filling its reservoirs. The only relatively permanent water bodies are the *aguadas*, human modified reservoirs that canalize rainfall and streams to deeper terrain. At present there are three large filled-up *aguadas* in Calakmul

The most important process that acts on surface materials (which are limestone, clays and vegetation cover), is the carbonate dissolution by rainwater. This kind of terrain has a predominant chemical action and subterraneous drainage characterized as karstic. In the Calakmul area, the rain falls and organic matter decomposes. Together, they produce large quantities of acid that dissolves the rocks. As a result, karstic valleys are often formed as closed depressions with subterraneous drainage. The *poljes* or large depressions, some of them extending over several square kilometers, have steep slopes and broad flat bottoms, sometimes with small residual hills and temporary or permanent lakes. The carbonate rocks have strata fractures that are important for the water currents, commonly creating caves and sinkholes (Gates, 1999).

Calakmul stands between Caribbean Sea and Gulf of Mexico watershed. Surface and subterranean drainage recharges the peninsular aquifers. The aquifer of the reserve is from the Paleocene, formed of gypsum and anhydrite in the 'Icaiche Formation'. It has depth of 60 to 165 m in the phreatic level (García Gil et al., 2002:52).

The Yucatan Peninsula as a whole is an open hydrological basin with a karstic aquifer of free type (Comisión Nacional del Agua, 1997). The high pluvial index and terrain elevation homogeneity permit a stable aquifer recharging. The direction of underground currents goes from eastern Campeche, the area of higher altitude where Calakmul is located, to the rest of the Peninsula in a radial direction. These characteristics permit a water-rich aquifer that has been supplying the Peninsula inhabitants for thousands of years. It has been confirmed that the contamination of the water with subsoil components as gypsum and anhydrites has a restriction factor for water consumption in the Calakmul region (Prado Roque, n. d.). However, Calakmul camp workers drink water from the big *aguada* everyday without becoming ill.

Regarding the subterraneous hydrology, the hilly zone is an area of consolidated material with high factors of water permeability (INEGI, 1989). This means that there is a high chance of a free type water-bearing mantle, filled from rainfall and which can be exploited through wells. The quality of the water is generally good. In contrast, the *bajo* area is a zone with unconsolidated material with low possibilities of water movement. This means that there are low possibilities of a water-bearing mantle.

- Vegetation

Calakmul Biosphere Reserve represents the largest forested area of the Mexican tropics. It has a large spatial heterogeneity, in spite of the relatively lack of topographic features (Martínez and Galindo-Leal, 2002:7). Thanks to multiple botanical studies that have been completed in the last decades, the characteristics of the Reserve biodiversity have been recognized (e.g. Flores, 1990; Martínez and Galindo-Leal, 2002; Ucán et al. 1999). Particularly, the vegetation of the elevated areas the site of Calakmul is considered as a middle semi-evergreen rainforest, with tree species as *chacah*, *zapote,* cedar, ramon and *guaya*. The vegetation of the lower areas (*bajo*) is classified as low semi-evergreen rainforest with species as *palo tinto*, black *chechen* and white *chechen.* The vegetation and soils will be described in more detail in the next chapters of this study.

- Fauna

A great diversity of fauna is found in Calakmul Biosphere Reserve. Eighteen species of fishes, sixteen amphibians and fifty reptiles, about tree hundred bird species and ninety-four species of wild mammals represent the biological diversity of Calakmul (INE, 2000). The most common species that have been observed around the site are listed in Table 3.1.

English Name	Scientific Name
Felines	
Panthera onca	jaguar*
Leopardus weidii	margay
Leopardus pardalis	ocelot
Herpailurus yagouaroundi	jaguarundi
Puma concolor	puma
Other mammals	
Urocyon cinereoargenteu	grey fox
Procyon lotor	raccoon
Potos flavus	kinkajou
Tayassu tajacu	collared peccary
Pecari tajacu	white-lipped peccary
Agouti paca	paca
Dasyprocta punctata	agouti
Nasua nasua	coati
Dasypus novemcinctus	armadillo
Tamandua mexicana	anteater
Tapirus bairdi	tapir*
Odocoileus virginianus	white-tailed deer
Mazama americana	brocket deer*
Ateles geoffroyi	spider monkey*
Alouatta pigra	howler monkey*
Birds	
Falco rufulgaris	black hawk*
Ortalis vetula	plain chachalaca*
Agriocharis ocellata	ocellated turkey*
Crax rubra	great curassow
Penelope purpurascens	crested guan
Amazona autumnalis	yellow cheeks parrot
Ramphastos sulfuratus	tucan*
Cathartes aura	aura buzzard
Reptiles	
Boa constrictor	boa
Ctenosaura similis	black iguana*
Crocodylus moreletti	crocodile
Micrurus diastema	coral snake*
Crotalus durissus	rattlesnake*

Table 3.1. Fauna of Calakmul. These are the most common animals near the site. The ones that I had the opportunity to see or that left an observable trace are marked with an asterisk.

Calakmul Archaeological Reconnaissance

Researchers have explored Calakmul and its periphery since the last century. Lundell was the first investigator who visited Calakmul, as well as other sites in the region (Lundell, 1933). In the same way, Ruppert and Denison explored Calakmul and some sites located in the nearby rainforest (Ruppert and Denison, 1943). In the 1980s, Morales López walked around many settlements located close to the recently paved road that goes to from the Escarcega-Chetumal highway to Calakmul (Morales López, 1985; 1987). He explored the first 20 km of this road and found about 25 archaeological sites. In 1999, archaeologist Javier López observed some landscape features located to the south of Calakmul (Omar Rodriguez 2003, personal communication). In 2000, a team headed by Agustín Anaya and Omar Rodríguez, from Calakmul Archaeological Project, explored the surroundings of Calakmul Great Plaza as far as 7.5 km to the south of the city. They also explored the Northeast and El Mercado groups (Carrasco et al., 2001). During this reconnaissance, they mapped the cultural features that they found.

Similarly, Folan and his team have done site reconnaissance in the periphery of Calakmul. They reported 56 archaeological sites in the road from Conhuas to Buenfil and Chumpich *Chicleros* Stations (Folan et al. 2001a). Some of the main sites are Oxpemul, Pared de Reyes, El Laberinto, San Felipe, Flor de Cacao and Uxul, probably occupied from the Late Preclassic to the Terminal Classic. In the last years, archaeologist Ivan Sprajc has also explored sites such as Oxpemul, Pared de Reyes, Uxul among others, near Calakmul (archaeologist Adrián Baker, member of Sprajc team, personal communication 2005).

Methodology for Studying Landscape

The first step for studying the landscape of Calakmul was to establish the landscape units through the "physiognomic-lithomorphic" (Sombroek and Van de Weg, 1980). A landscape unit is an area of distinctive relief, geography and soil elements (Breimer et al., 1986). First, landforms are described considering physiography and altitude, for example, uplands (U) or dunes (D). Each landform should have similar physical characteristics and hence a uniform origin. Second, the basic geological information is added to the landform legend, as for example MF, mountains (M) with rocks rich in (F) ferromagnesian minerals (Breimer et al., 1986).

The proposal of this research was to undertake fieldwork to compile information at a 1:20 000 scale. This more detailed scale is useful for assessing the potential for agricultural development and also susceptibility of soils to deterioration. In addition, it can demonstrate suitability of soils for management and land preparation, constraints to land use, and it is practical for the planning of any land settlement scheme (Breimer et al., 1986). This method has been employed for establishing setting characteristics in developing countries. In particular, UNESCO has used it for tropical soils development projects employing the FAO soil classification.

The base image for performing the spatial analyses was the ortho-photograph of Calakmul, scale 1:20 000 (INEGI 2001b) and the similar 1:20 000 vector digital topographic map of the area (SIGSA, 2004). An ortho-photograph is an aerial photograph which landscape and camera lens deformation is corrected. Hence, the elements observed in the landscape have the same scale and it is possible to measure precisely distance and areas. In this study, INEGI digital ortho-photographs E16A81b and E16A81e were used.

In addition, an ASTER image was used for observing vegetation differences through its several color bands. An ASTER radiometer (Advanced Spaceborne Thermal Emission and Reflection Radiometer) is an imaging instrument flying on Terra, a satellite launched in December 1999 as part of NASA's Earth Observing System (EOS). ASTER is being used to obtain detailed maps of land surface temperature, reflectance and elevation (http://asterweb.jpl.nasa.gov/). It is designed to obtain high-resolution global, regional and local images of Earth in fourteen color bands.

Additionally, the map of Calakmul elaborated by May Hau and his colleagues was employed for further interpretations. This map was elaborated from 1982 to 1989, when Jacinto May Hau, Francisco May Hau, Nicolás Caamal, Cándido Choc and Rogerio Couoh mapped thirty km^2 of the site with a Brunton compass and some times with a theodolite. They established a grid of 120 squares of 500 m^2 each one and mapped cultural and natural features (Folan et al., 2001a). The information was re-drawn by Gonzalez Heredia whom transformed the data to a 1:2000 scale. Since this map shows different kinds of ancient landform manipulations (i.e. quarries, reservoirs) it is very useful for archaeological research. However, it was not possible for me to fit this map with the ones of INEGI, SIGSA or raster images since it does not have a projection and it seems that the deformation error is large and uneven. As a consequence, the map could not be geo-referenced and the spatial features could not be measured with confidence. However, it was used as an additional reference for making interpretations.

Calakmul Landscape Units

There are four major landscape units in Calakmul periphery: the structural plateau of karstic development, the strong and moderate hillside slopes, the eroded relief and the karstic depression (called in Maya *akalche*; see Figure 3.6).

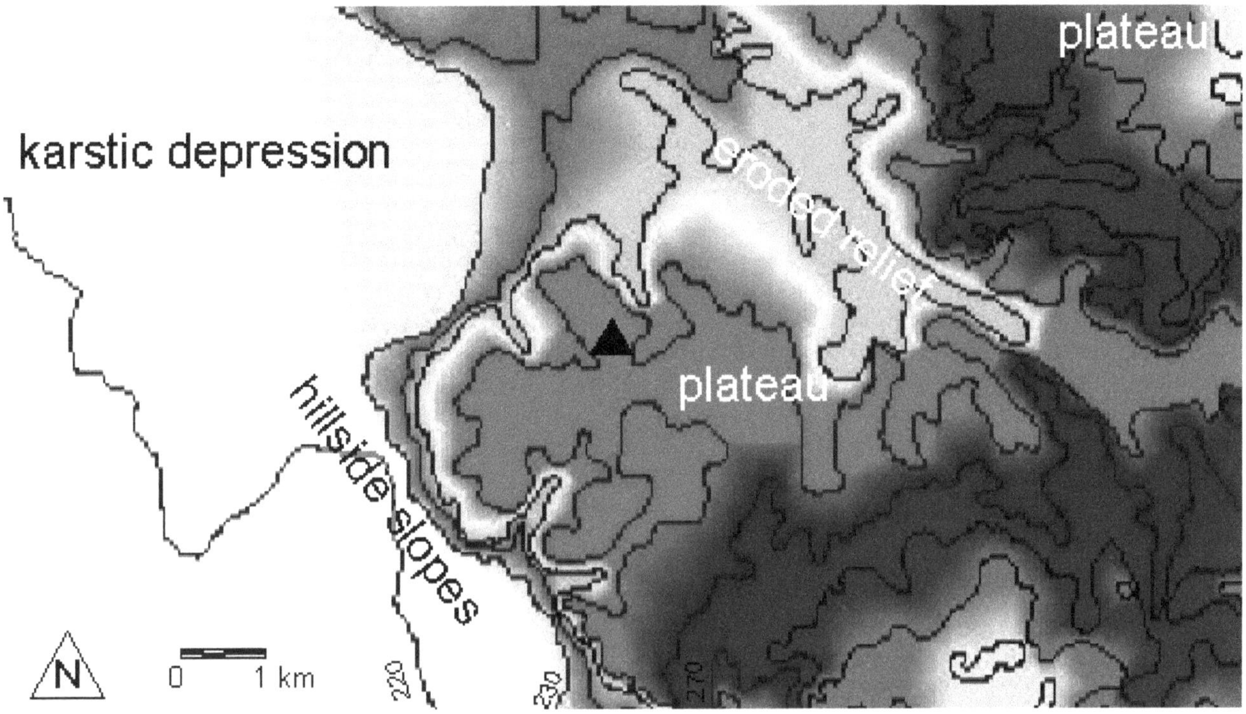

Figure 3.6. Map of the landforms of Calakmul. This map is based in a digital elevation model of Calakmul, scale 1:20 000. The distance of each contour line is 10 m. The triangle indicates the location of Calakmul civic-ceremonial center; the scale measures 1 km.

- Structural Plateau of Karstic Development (PL)

The site of Calakmul is located in a karstic plateau with a northwest-southeast direction that evolved from a concavity of the limestone platform that reaches 380 meters above the sea level (masl; García-Gil et al., 2002). The hills reach between 20 and 50 meters in height and are located between 250 and 380 masl. Between the hills are micro-valleys. It is relatively easy to walk in these areas. The hills have multiple non-permanent streams as well as sinkholes. The soils are well-drained, shallow, with grayish brown colors, friable and have developed above highly calcareous rock or soft parent material with a silty clay loam texture. The predominant vegetation is medium rainforest. Several non-permanent streams pass through the hills channeling all the water to the plains that ultimately reaches the karstic depressions.

A tri-lobed prominence stands out from the series of northwest to southeast hills that form the plateau (See Figure 3.7). The prominence where Calakmul site is located is divided by three steep ravines, two of them located to the south, and one to the northwest. Several micro-valleys are found in this prominence where a high number of remains have been mapped.

- Strong and Moderate Hillside Slopes (HsL)

The nature of the karstic landscape does not allow well-developed hillside slopes, but still in the margins between plateaus and depressions they can be observed. The edge of the prominence where Calakmul is located has a very steep relief in all directions except for the north. Based on the observations of García-Gil and his colleagues, very steep hillsides (more than 25°) with deep ravines and moderate hillsides (between 6° and 25°) are present (García-Gil et al., 2002). These landscape units have very shallow soils located over hard widespread rock outcrops or weathered limestone. They have a black or very dark grayish brown color A horizon (topsoil), weak structure, good drainage, abundant stones and a sandy loam texture. Limestone outcrops were present next to these soils. Also gypsum outcrops were detected.

- Eroded Relief (ErL)

The water flux that runs from the elevated areas to the karstic depression is channeled through the dissected eroded areas. These areas located to the north of the site of Calakmul, are inundated seasonally. The rainwater then drains slowly toward *Bajo* el Laberinto. These areas alternate with small hilly areas and it seems that some earthworks (ditches and channels) were made in order to impede water movement or to capture water or humidity. These zones have black to brownish gray soils with surface cracking, poor or moderate drainage, moderately deep soils depth and have a clay loam texture.

- Karstic Depression (KdL)

Considered also as a valley of karstic development, the *Bajo* El Laberinto depression is crossed by an intermittent current called the Laberinto River and subjected to seasonal inundations (García-Gil et al., 2002). This huge elongated flat area is located between 220 and 230 masl. It has a perimeter of approximately 143.83 km and an area of 391.5 km^2 (measure based on an ESRI EarthSat image, 2004). The slope degree of the area ranges from approximately 1 to 2%. This area is flooded during the rainy season. All the water that is captured from the plateau is directed toward a geological fracture with a northwest-southeast direction that forms the bottom of *Bajo* El Laberinto.

The Boundaries of Calakmul

Estimating and delineating the boundaries of sites as they were for ancient inhabitants is no easy task for archaeologists. The 1:6250 Calakmul map demonstrates that the site is satiated with ancient structures, except in the *bajo* area. Multiple structures have been reported in contiguous areas. Lundell observed innumerable scattered house mounds in the uplands of northeastern Peten (Lundell, 1937). The presence of 25 sites reported by Morales López in the first 20 km of the road to Calakmul suggests that at least this region was densely populated in ancient times, since almost every suitable space was occupied (Morales López, 1985; 1987). For these reasons, the proposal of specific site limits or boundaries in this area must be managed with care.

Throughout human history, landscape attributes have been significant elements in establishing borders and frontiers. Features like mountains, rivers, lakes, and swamps are natural barriers that impede entry and define boundaries. Similarly, cultural features such as moats (for example, at Becan, about 80 kilometers from Calakmul), walls and trenches have been constructed to restrict access. Hence, the features of landscape can be evaluated in order to surmise their relevance in terms of the delineation or circumscription of particular areas.

Considering the naturally occurring environmental restrictions, the landscape features that most probably formed the natural boundaries of the site of Calakmul were ones that surround the promontory: the eastern and southeastern hills, the peripheral seasonal stream (or *corriental*), the seasonally inundated areas located to the north of the site, and the *bajo* El Laberinto to the west. All of these are practical, viable boundaries that are further enhanced during the rainy season.

The city is bordered on the north, west and south by the *akalche*. During the rainy season upland terrain is flooded, but since the seasonal streams channel water to lower areas, this inundation does not last long. In addition, the streams and channels that were observed could have canalized as much extra water as possible, further preventing long-term flooding. As a result, water flows to lower sections (*akalche*) where it can remain for months, impeding human access to large areas.

The *corriental*, a non-permanent seasonal stream that surrounds almost the entire promontory of Calakmul, is not uniform. In the hilly area, it is deep and thin, and enlarges near the largest *aguada* in the north. In some areas the riverbed is about one and a half meters of width and in others, it extends up to five meters. Clearly, humans modified this feature to channel available water. It is also possible to appreciate features that resemble ditches, and that appear to have been used to store water. This fact will be discussed below.

The eastern hills, with heights of about 270 masl near Calakmul and 380 masl at their summits, also form a natural boundary, though not an impenetrable one. The steep topography, small ravines, seasonal riverbeds, lack of surface water, and thin soils certainly impede human occupation in the hills, but do not definitively preclude it. May Hau and his colleagues (Folan et al., 2001a) have mapped some scattered ruins in these areas.

As in other Maya cities, such as Bonampak or Yaxchilán, the inhabitants of Calakmul took advantage of the natural elevations present at the site for the construction of their buildings. In Calakmul principal structures and palaces were distributed in an elliptical area of approximately 1.4 km^2. In contrast, platforms grouped as plazas and scattered minor structures are distributed over the entire site, except in the *bajo* area. These structures were most likely the houses of the commoners. May Hau and colleagues did map some platforms in the *bajo* area but they proposed that these structures were used in the process of exploiting flint from the *bajos* rather than for inhabitation (Folan et al., 2001a).

Hydrology of Calakmul

The whole of the Yucatan peninsula is a karstic platform. The term karst is applied to a landscape that has been modeled by the dissolution of limestone rocks by water. An important part of a karstic landscape is its hydrology, which is primarily subterraneous, although there are some formations that retain surface water, such as sinkholes or the low-developed intermittent fluvial system. This system is commonly dendritic or sub-dendritic, with a pattern that resembles the branching of a deciduous tree or a neuron dendrite.

A careful study of the site of Calakmul reveals the ways in which its ancient inhabitants addressed the problems associated with the procurement of water. The landscape elements related to water in Calakmul are: *bajos*, *aguadas*, sinkholes, *corriental*, seasonal water channels and *chultuno'ob*. Other features as *sartenejas* and limestone quarries are also discussed as potential water reservoirs (Figure 3.8).

- Bajo

In its upper reaches, El Laberinto receives water from several non-permanent streams, two of which are located to the east and west that canalize rainwater in large quantities. The streams arrive at the *bajo* channeling water that finally moves to the bottom of the depression. Interestingly, the upper part of El Laberinto forms a rounded pool bounded by the stream that originates in the *corriental* of Calakmul. Similarly, the stream that originates in the largest gully of the Calakmul promontory forms the border of what seems to be a more elevated area compared to the pool just mentioned. This is recognizable due to the fact that this zone floods less during the rainy season, in comparison with other areas in the immediate vicinity. In other words, it seems that the terrain is canalized distinctly when compared to adjacent areas that exhibit a greater level of inundation. If El Laberinto *Bajo* is visualized as a mushroom, the cap and stem have a greater level of inundation if compared with the top of the stem (Figure 3.2).

The vegetation of the *bajo* has been called *akalche*. The most common species are described in Chapter 4 of this study. It is remarkable how difficult it is to navigate this kind of forest since lianas, logwood (*palo tinto* trees) and fallen vegetation combine to form an intricate mesh. This vegetation is subjected to extreme hydrological changes, alternately dry and waterlogged for several months on end, depending on the season.

Cooke (1931, cited in Lundell 1937:5) believes that *bajos* once formed shallow lakes, which were silted by the erosion of the uplands until they lay above ground-water level during the dry season. Heavy clays prevented easy sub-surface, and hence, extended swampy areas were formed.

When we walked through the *akalche*, there were some features similar to the *cuyos* (small terrain undulations) that Folan and Hansen described and subsequently proposed as a viable agricultural matrix (Folan et al. 2001a:26; Hansen et al., 2002). If the *cuyos* they report are similar to the ones we saw, I differ with respect to the agricultural viability they suggest. In El Laberinto, undulations are the result of soil hardening around plant roots. For example, as *palo tinto* grows, it traps soil within its roots and with time the matrix becomes harder than in places where soil flows more easily as a consequence of inundations. When the plant dies, the soil stays hard and the *cuyo* is formed. One can observe in *akalche* that these *cuyos* are formally irregular; some are quite large, while others are very small.

- Aguadas and Sinkholes

An *aguada* or doline is a small body of water formed from an ancient sinkhole called *cenote* in Spanish and *ts'onot* or *dzonot or tzonot* in Maya. As a result of erosion, the vertical sinkhole vertical walls breakdown and its bottom gets filled with nearly impermeable clays and organic matter. This phenomenon breaks off the connection with subterranean waters in some cases (González Medrano, 2003). In others, *aguadas* maintain a connection with an under-drainage system and retain water, but often for only a few hours. Many are present only during the periods of extended rain. *Aguadas* modified by humans have been identified in many archaeological sites in the Maya lowlands. For example, in La Milpa, Belize, an *aguada* was modified to improve water retention and one of its margins appears to have been deliberately raised (Scarborough et al,. 1995:112).
There are two different types of *aguadas* in Calakmul. The first is comprised of deep, permanent reservoirs that contain water year-round. We observed three large *aguadas* in the site. These are the biggest reservoirs of the site and its surroundings, located at the lowest elevations, and with several seasonal streams and channels feeding into them. Moreover, they seem to be connected with other watercourses as well. It appears that the principal *aguada* in Calakmul is the largest yet encountered in the Maya area (Martin and Grube, 2002). In the northern Yucatan, researchers have reported that the quality of water contained in permanent *aguadas* is extremely low, with anoxic hypolimnetic waters and high levels of ammonia and H_2S, which limit fish production (Flores Nava, 1994). However, there was a variety of fish in all the *aguadas*, including some quite large specimens in the principal one.

The other type of *aguada* consists of seasonal basins, only recognizable by form during the dry season. These depressions are filled with rainwater during the wet season. On the Yucatan Peninsula nutrients retained in the soil and vegetation, such as phosphorus and nitrogen, are released in the water with the rains. Then, when the *aguada* dries up, the nutrients are readily available for cultivating. It is important to note that these water bodies are very rarely used as sources of potable water in northern Yucatan (Flores Nava, 1994). Sinkholes were only detected from aerial photographs in the periphery of the site.

After examining aerial photographs, satellite imagery, and in the field, an intricate pattern of streams that channel water into reservoirs appeared, but it is difficult to differentiate from remote sensing images if it is natural or artificial. The idea of the existence of channels in Calakmul has already been suggested (Adams, 1981 personal communication in Folan et al., 2001a:26). In this study their locations and plausible functions will be interpreted.

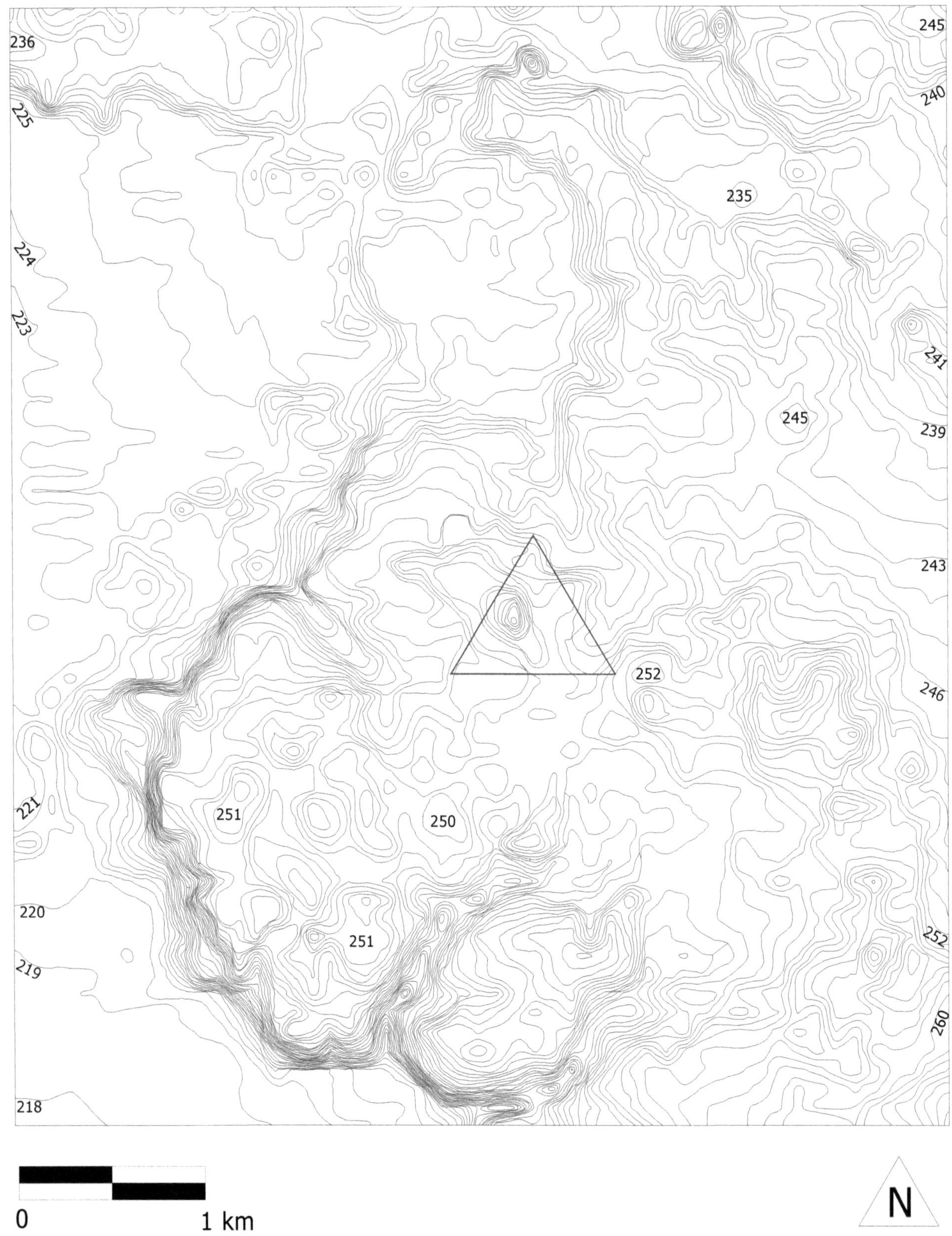

Figure 3.7. Topographic map of Calakmul. Redrawn from Folan and colleagues (2001a) and May Hau and colleagues (1990). The quantities indicate meters above the sea level (masl). The triangle represents the site's civic-ceremonial center.

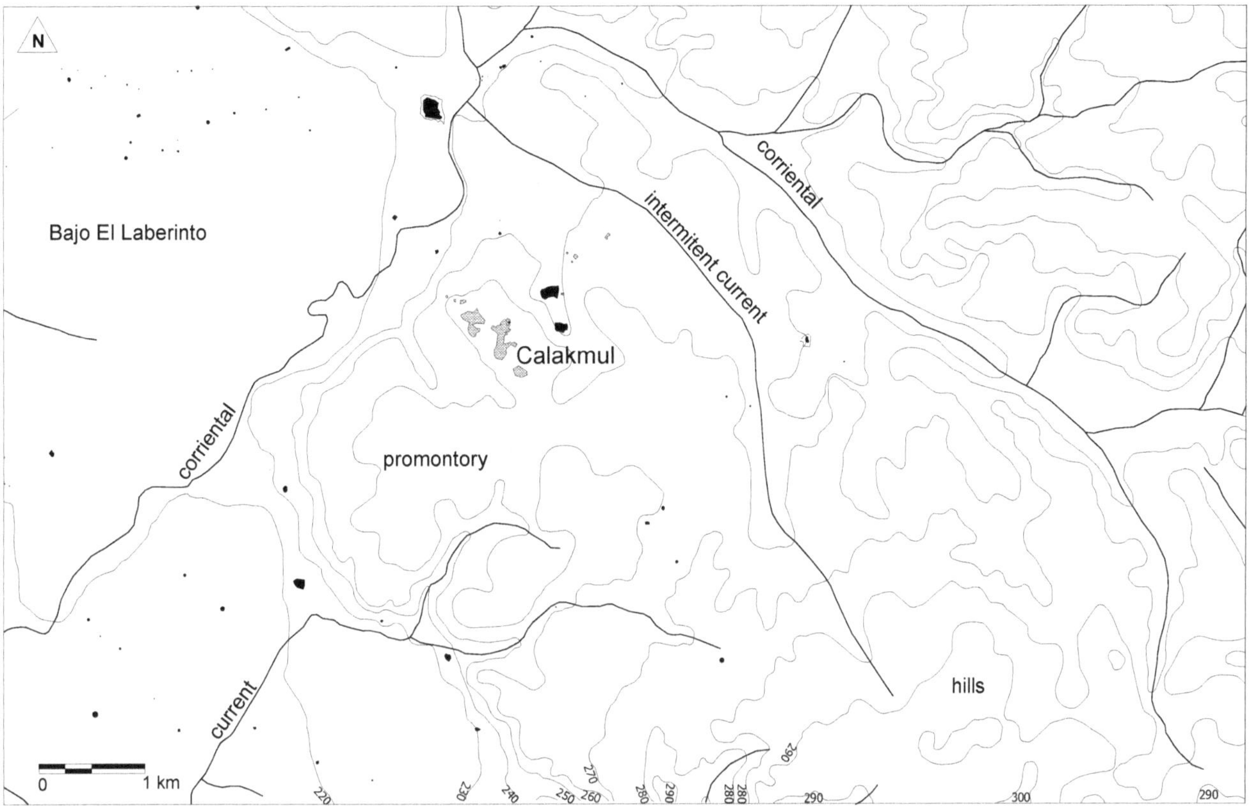

Figure 3.8. Hydrological features of Calakmul and periphery, such as intermittent streams, water reservoirs and the Bajo El Laberinto (Hydrological information based in SIGSA, 2004). The light gray colored area represents the site's civic-ceremonial center. The reservoirs are represented in black.

- Corriental

A non-permanent seasonal stream, the previously described *corriental*, surrounds the entire promontory. It is quite very possible that the Maya modified this stream, due to the fact that artificial features such as reservoirs appear in its drainage basin.

- Non-Permanent Streams

Several kilometers of non-permanent streams, drain into *aguadas*. It seems that the Maya developed a complex pluvial web to take advantage of the heavy rains. Many of these streams form lattices in areas of *akalche* vegetation.

- Chultuno'ob

An interesting feature found in Calakmul is the *chultun* (*chultuno'ob* in plural) or subsoil reservoir. These human-built cisterns are underground chambers with a mouth-like opening and long neck that expand at the bottom in the form of a boot. In Calakmul there are single and double chamber *chultuno'ob*. Zapata excavated and described five of them (1985), and proposed that they could be used for water or food storage, or even as dumps.

- Sartenejas

Another feature is the *sarteneja* or small cavity carved directly in the rock, probably used as a water container. Gallegos (1985) reports two of them near a quarry she excavated, and concludes that they are in some way related to the process of stone extraction.

- Limestone Quarries

Quarries have previously been reported in Calakmul (Gallegos, 1985), and their locations have been reported before in the 1:6250 map (Folan et al., 2001a). We were able to observe few of them. Interestingly, some of these quarries are deep conical depressions located next to architectural complexes. There is a significant possibility that these depressions were excavated quarries that, intentionally or not, were later converted into temporary reservoirs. This prospect can only be confirmed through excavation. As a reference, in northern Yucatan many contemporary quarries rapidly fill with sediments due to the shallow water table, and are subsequently used by farmers for aquaculture (Flores Nava, 1994).

The Channel System of Calakmul

Interestingly, the presence of multiple streams and water channels were detected (Figure 3.9). Many of them have formed principally in areas with non-plain slope for canalizing the rainwater. There were also some currents in nearly flat areas; for example the large stream that unites the two permanent *aguadas* located to the east of the civic-ceremonial center.

The presence and plausibility of channels in Calakmul has been previously reported and examined. Folan reports that Adams observed a lattice in the Laberinto *bajo* in radar images obtained from NASA (Folan et al., 2001a:26). Along similar lines, Dominguez Carrasco has excavated a kind of drainage (she reports that it is a channel with a bridge) between the nearby *aguadas* (*Aguadas* No. 3 and 4; Folan et al. 2001a).

After performing an analysis of remote sensing images, a complex network of streams and channels appeared (Figure 3.9). Some of them can be recognized from Landsat images but the majority is visible through ortho-photographs. After closely observing the area there were features such as the ones shown in figures 3.10 to 3.13.

The majority of the water channels are located in the transitional area between *bajos* and elevated areas, particularly in the northern terrain dissected by erosion and in the *bajo* that surrounds the promontory. Some of the features may be natural streams but others form what seem to be artificial reservoirs and lattices. Channels are connected to main streams and probably canalized water to specific fields. In Calakmul, groups of streams form lattices in different regions within *bajos*. Each lattice unit (the polygon) has a variable area that ranges between 1300 to 5500 m^2. These channels could have supplied water to fields and also drained them. The excess water then could have been canalized to the multiple reservoirs simply through the force of gravity

Gravity fed channels, as well as water manipulation systems that implied channeling water into reservoirs, have already been reported in Mesoamerica (Scarborough et al., 1994; Scarborough and Gallopin, 1991). The channel system of Edzna in Campeche is a prime example of the complexity of water management activities undertaken by ancient Maya kingdoms. In this site, a channel network guided water to a reservoir and at the same time was utilized for field irrigation (Matheny, 1976). Siemens and Puleston found these features in the Candelaria River in Campeche (Siemens and Puleston, 1972). Turner and Harrison (1981) report a complex channel network in Pulltrouser Swamp in Belize. Channels have also been detected in Cerros, Belize (Freidel, 1982).

Although from aerial photographs and remote sensing images it is difficult to distinguish which features are artificial, the pattern that I observed in Calakmul makes me believe that the Maya modified the streams in order to direct water at their convenience.

***Corriental* Water Reservoirs**

Another interesting feature is what it has been labeled as *corriental* water reservoirs (See Figure 3.10). It seems that the meanders of the *corriental* were used to direct water into reservoirs. A clear pattern that repeats almost through the entire stream is a concavity located on the internal perimeter of the meander connected by one or more small channels to the main one. This feature may have served to catch water and hold it for longer periods. In fact, Domínguez-Carrasco and Folan report that one of these reservoirs is an *aguada* (*Aguada* No. 2), which seems a distinct possibility (Domínguez Carrasco y Folan, 1996). But due to continued sedimentation, it is not as deep as it would have been before, and hence does not retain water as well as other major *aguadas* in Calakmul.

A further possibility is that these patches of land were irrigated for intensive cultivation, as has been observed in La Milpa, Belize. At that site, reservoirs were used for regulating the moisture levels of soil in flat, upland pockets of highly fertile soils, which appear to have been used as plots suitable for intensive agriculture (Dunning et al., 1999).

Circular Feature

There exists an interesting circular feature that can be observed in satellite images and aerial photographs. Although it was not possible to directly confirm its existence in the field, it appears to be associated with the redistribution of water from higher to lower terrain. It could be a mechanism by which excess rainwater was retained and distributed. Since this feature is located at a lower altitude and linked to previously identified potential reservoirs located on higher ground, it is probable that this feature concentrated and channelled the water from the adjacent hills. Field reconnaissance and excavation are necessary in order to discern the function of this feature, which can be observed in Figure 3.11.

Lowering the Water Flux

A series of potential hydraulic features can also be appreciated through remote sensing images. Since abundant rain can inundate the terrain in short time, some landscape features can be used to lower the water flux. In Calakmul, the *corriental* confines the water in adjacent areas. The *corriental* runs from southwest to northeast into *bajo* areas, catching the water on one hand, diverting it and avoiding water excess loss through infiltration and evaporation on the other (See Figure 3.12). In the same way some features that resemble ditches or channels, can also be useful for retaining water and canalizing it (See Figure 3.13).

These water-related features were detected in aerial photographs and ortho-photographs. At field, several streams all over the place were also observed. Some of these features may not have the function that has been mentioned. However, it is plausible that they were modified by humans for collecting and redirecting rainwater. Hence, they are probably related with landscape management. Future excavations will confirm the function of these land modifications.

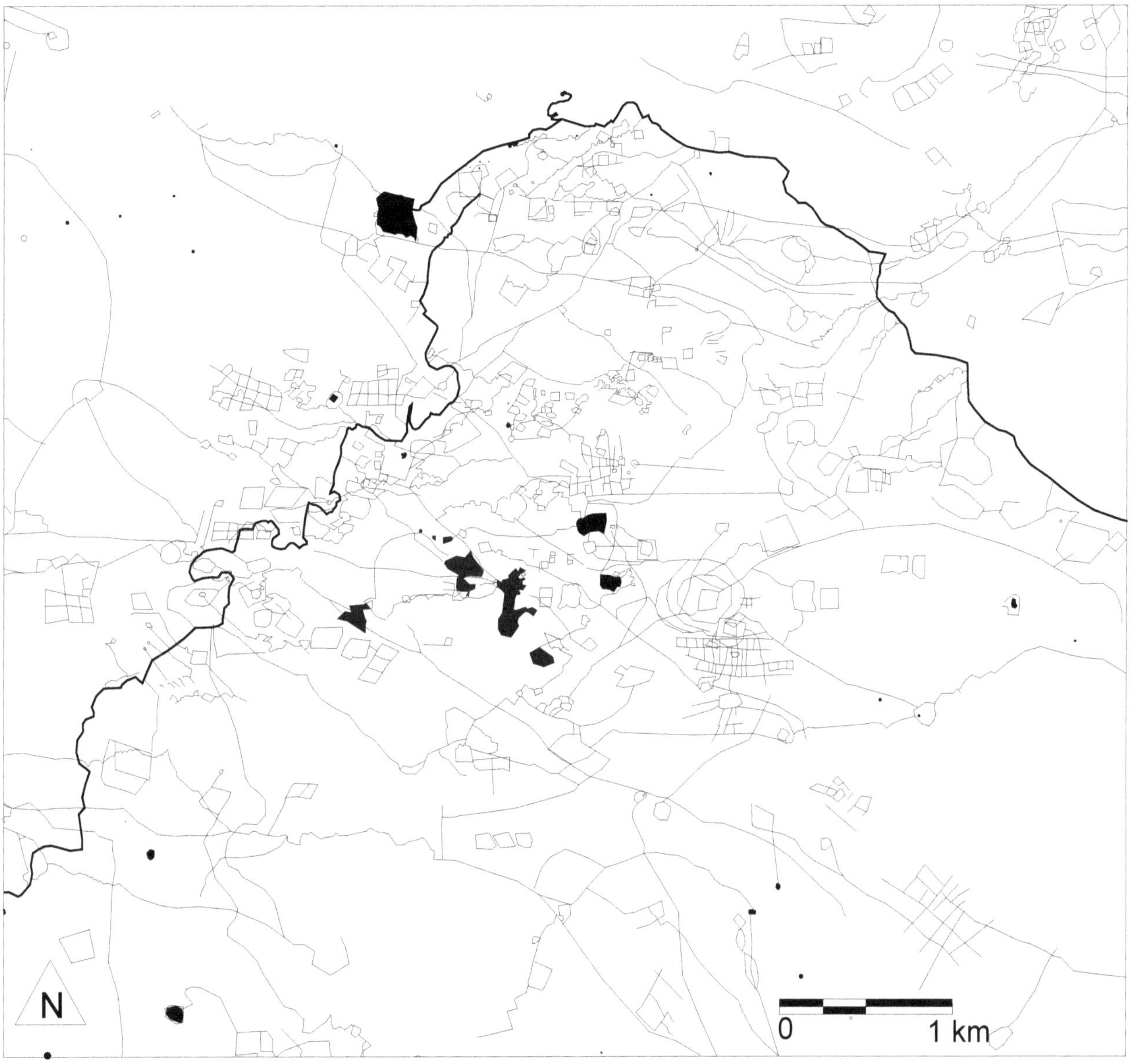

Figure 3.9. The complex network of streams in Calakmul. The coriental, channels and streams are shown in light gray; site's civic-ceremonial center in dark gray, and reservoirs in black.

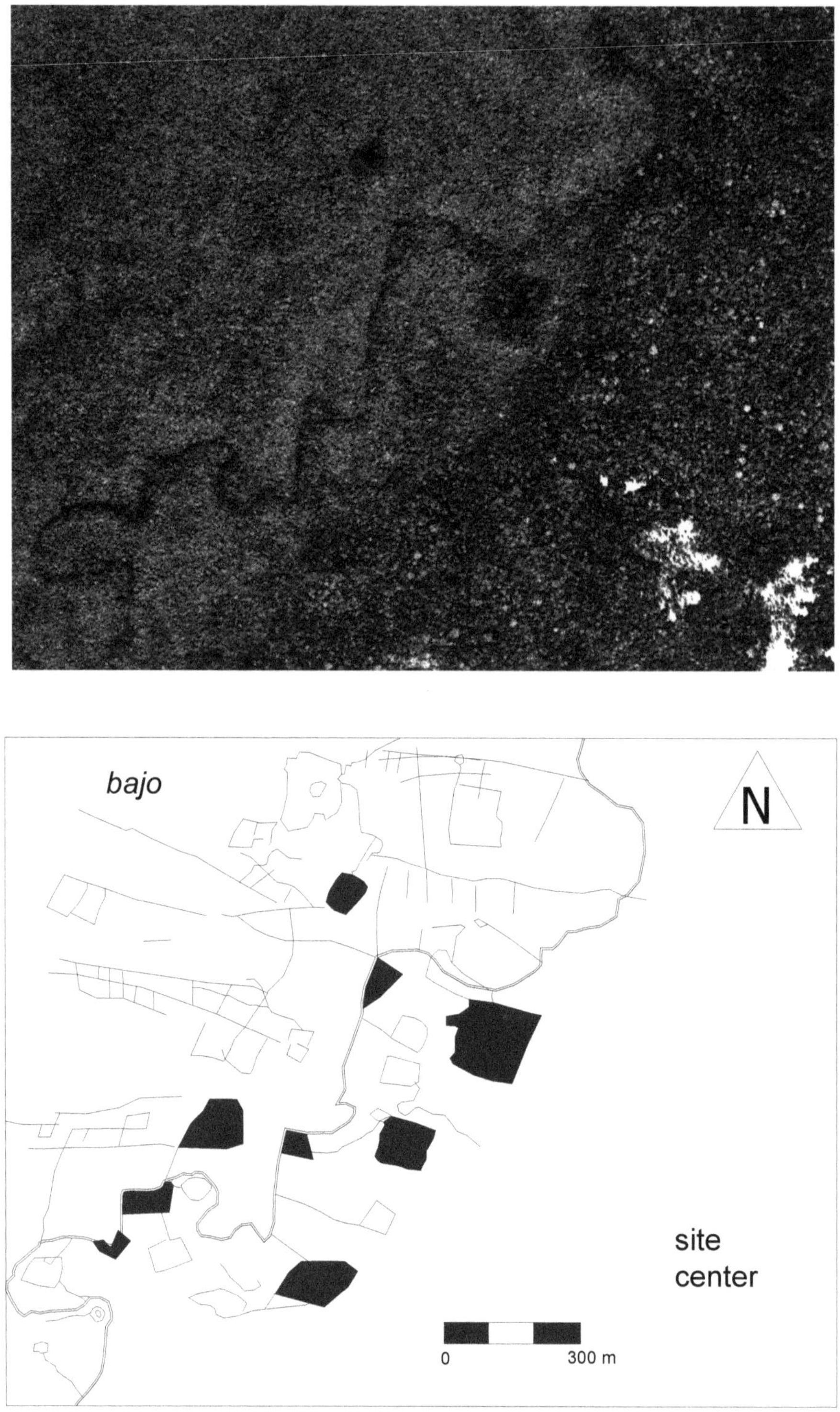

Figure 3.10. Calakmul corriental reservoirs. The superior image is an ortho-photograph 1:20 000 (INEGI, 2001b) showing the corriental with adjacent hydrological features. The inferior image demonstrates the same features on the same scale in a vector mode. The reservoirs are shown in black; corriental and potential channels in grey.

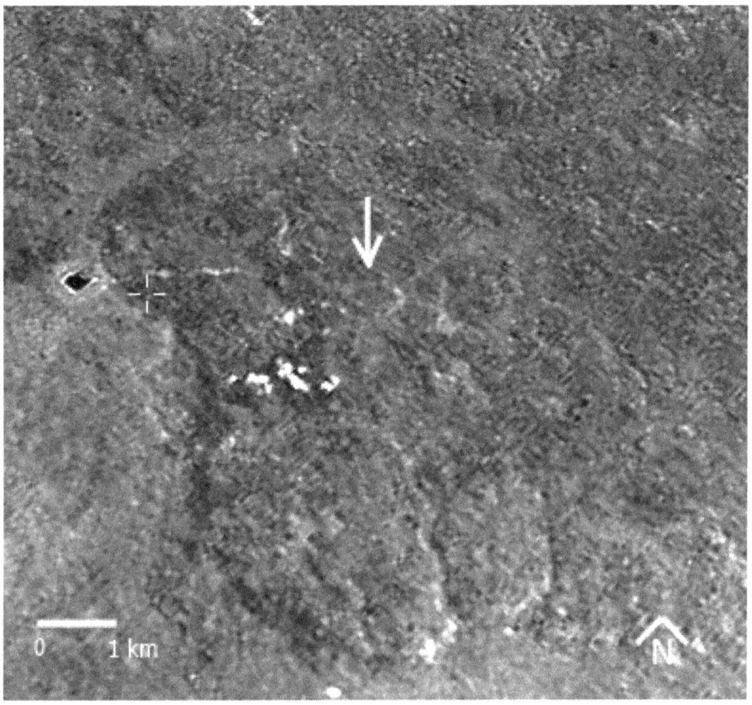

Figure 3.11. 3D remote sensing image showing a circular feature (pointed by the arrow) with four adjacent lines features that converge in it, probably channels. This feature is situated to the east of the largest aguada (the square in black) of Calakmul.

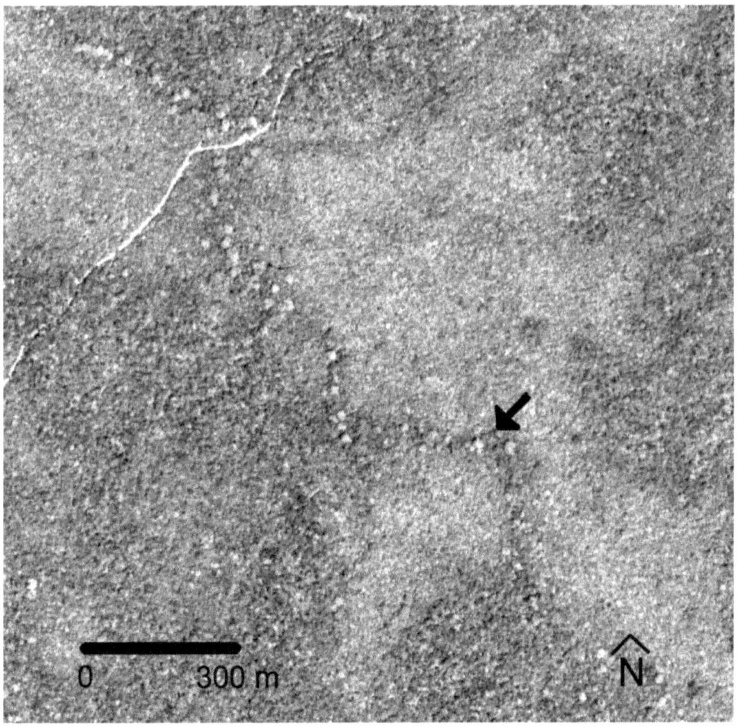

Figure 3.12. Aerial photograph of the corriental in the northeastern part of the Calakmul promontory (INEGI, 2001a). The white line in the left upper part of the image is the main road to the archaeological site. The arrow points to a section of the corriental possibly modified to catch and divert flowing water from akalche area.

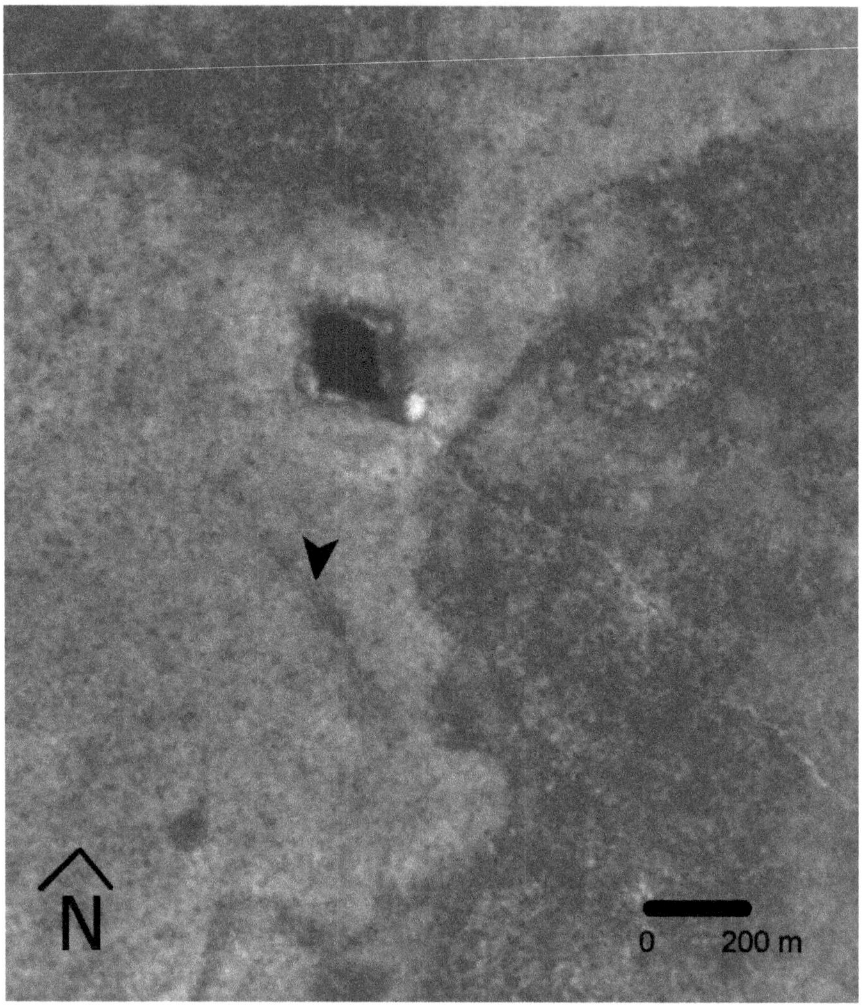

Figure 3.13. Ortho-photograph 1:20 000 (INEGI, 2001b) of a linear feature (pointed by the arrow), that resembles a ditch or a channel.

CHAPTER 4. THE VEGETATION OF CALAKMUL

"Even though no actual written records exist giving accounts of the practices of the ancient Maya, survivals among the Indians of today, evidences from archaeological investigations, and studies on the distribution of species of plants give important clues. The distribution of certain species can be explained satisfactorily by taking into account the rôle which they have played among the Maya."

Lundell, 1937:10.

The Vegetation of Calakmul and Periphery

Formal studies of the Calakmul Biosphere Reserve vegetation have been made since the 1930s. Lundell surveyed Calakmul area and identified the species for the first time (Lundell, 1934). During the 1950s, Faustino Miranda collected numerous plant materials and published the results (Miranda, 1958). During the 1980s, Edgar Cabrera Cano and Edilberto Ucán also collected plant specimens (Martínez et. al., 2001). From the 1990s and until today, the National Autonomous University of Campeche and the National Commission for Knowledge and Biodiversity Use began an inventory of flora in Calakmul region that has documented more than 1000 species, samples of which are currently kept at University of Campeche. In addition, 370 species have been added as part of a project orchestrated by Calakmul Reserve management team, Xpujil Regional Counsel and International Program for Model Forest (Martínez et al., 2001). Recently, scholars have identified, described and classified the flora of Calakmul Biosphere Reserve and its distribution (Díaz-Gallegos et al., 2002; Flores, 1990; Gutiérrez, 2000; Martínez, 1999; Martínez et al. 1999, 2001; Martínez and Galindo-Leal, 2002; Ucán et al., 1999).

The vegetation diversity is an attribute that stands out in the Calakmul Biosphere Reserve. In general terms, it can be classified as a tropical wet forest. The trees reach about 30 m, with lianas and epiphytes in their trunks, small palms and an almost open forest floor in elevated areas. This kind of forest has been called true rainforest and occurs in Alta Verapaz and Peten, Guatemala, south of Belize, south of Campeche, east of Chiapas and parts of northern Honduras (Schlesinger, 2004).

After exploring Calakmul Biosphere Reserve during the 1990s, Martínez and Galindo-Leal observed six vegetation types: High evergreen forest, medium and low semi-evergreen forest, low deciduous forest, savanna and hydrophytes: *popal*, *carrizal* and *tular* (Martínez and Galindo-Leal, 2002). They generated a list of 1569 taxa including 147 families, 726 genres, 1537 species, 5 subspecies, 23 varieties and 4 forms (Martínez et al., 2001).

In the site of Calakmul and its periphery it is immediately possible to make two observations through remote sensing images. First, the area is almost totally covered by vegetation (Figure 4.1). Second, there are two major types of vegetation. A subsequent reconnaissance of the site confirmed that another type of vegetation was associated with reservoirs. The contrast between two major types of rainforest vegetation can be observed in the INEGI aerial photograph, scale 1:75 000 (INEGI, 2001a; Figure 4.2).

The vegetation in Calakmul and its periphery has been classified as medium and low semi-evergreen forest (INEGI, 2002). Four strata of vegetation, as seen in Figure 4.3, characterize medium semi-evergreen forest. The vegetation looses between 25 to 50% of the leaves during the dry season. This forest is prevalent in the hills. The tallest trees, with a height of 15 to 30 meters are found in the upper stratum. In these trees are epiphytes as bromeliads and orchids and abundant lianas. The next lower stratum presents trees with height between 7 and 12 meters, and the next stratum down has plants between 1 and 4 meters. Finally, the lowest stratum has few plants (some of them trees in a developing stage), many leaves, plant fragments and an open floor.

The species of each of the strata are mentioned as part of Table 4.1. This listing was obtained from INEGI sample points 78, 81, 88 located in the site and its periphery (INEGI, 2002), and from Lundell (1937). The asterisks indicate the species that were observed in the field.

The low semi-evergreen forest is found in the seasonally inundated flat wooded swamps called *bajos* in Spanish or *akalche* in Maya. The stagnant water contains a high percentage of decomposing organic matter. In some months the surface water completely evaporates creating extreme conditions for plant survival. The *bajos* clearly present two strata of vegetation, one with species up to 10 meters of height and the other with plants with 2 meters of height (See Figure 4.4). The plants lose from 25 to 50% of their leaves during the dry season. In El Laberinto *bajo* large quantities of *palo tinto* (*Haematoxylum campechianum*) can be observed.

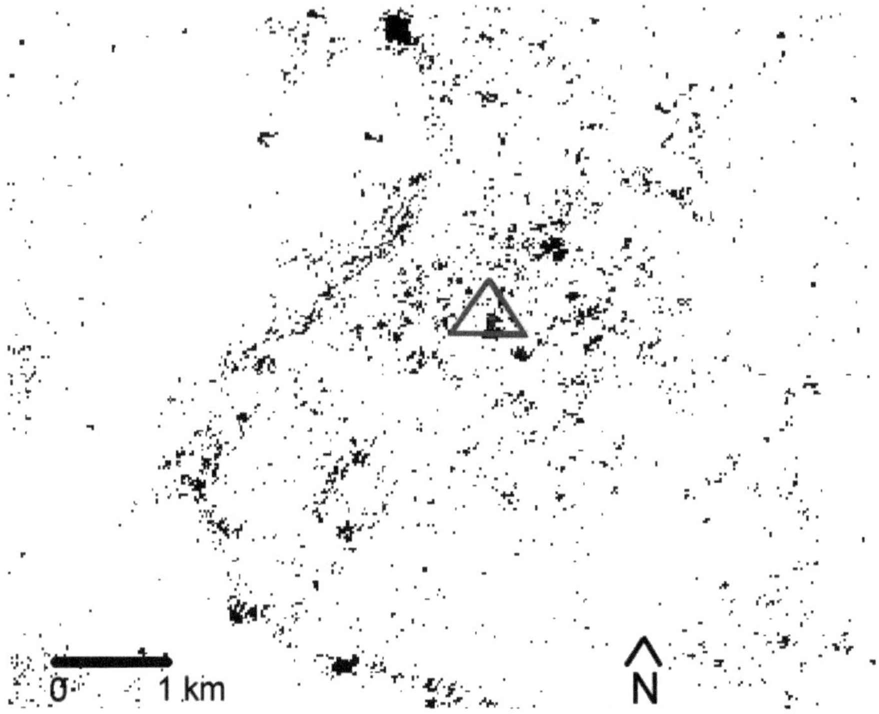

Figure 4.1. Aster image of Calakmul. Unsupervised classification of Vnir 61134a band. The area with vegetation is shown in white; the surface without vegetation, in black. The civic-ceremonial center is denoted by the triangle.

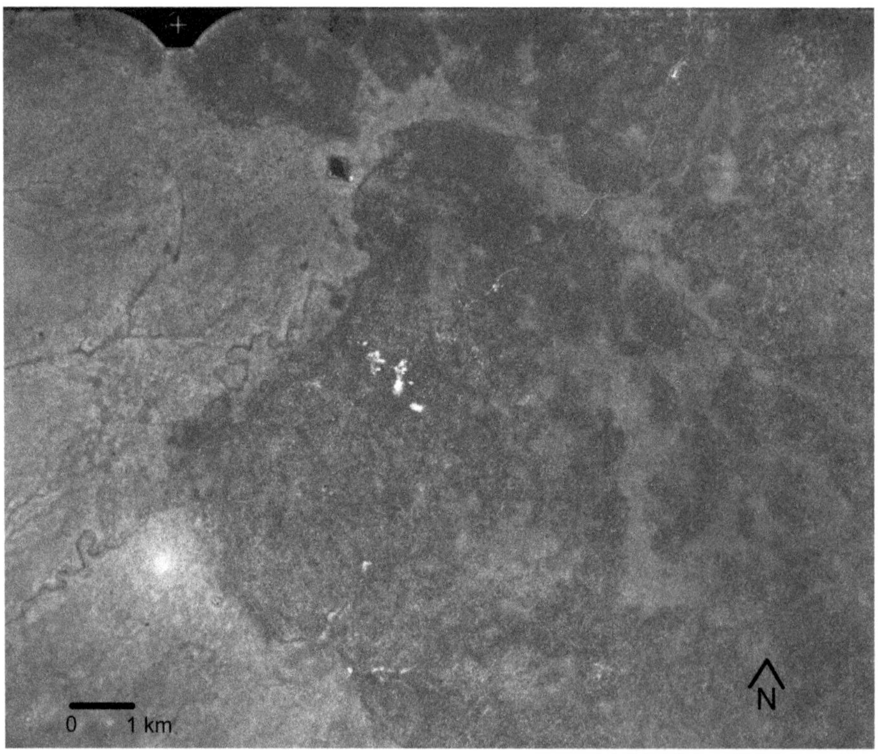

Figure 4.2. Aerial photograph of Calakmul and its periphery showing the difference between two types of vegetation (1:75 000; INEGI, 2001a). The site's main buildings appear in the center of the image in white color. The biggest aguada (water reservoir) is located to the north of the site. Notice the difference between low semi-evergreen rainforest colored in light gray and medium semi-evergreen rainforest in dark gray.

Scientific name	Spanish name	Maya name
Stratum between 15 and 30 m high:		
*Manilkara zapota**	*zapote, chicozapote, chicle*	*ya´*
*Brosimum alicastrum**	*ramón*	*chak óox*
*Bursera simaruba**	*palo mulato*	*chakah*
Pseudobombax ellipticum	*amapola, clavelilla*	*xkuxche'*
*Talisia oliviformis**	*guaya, guayo*	*uayum*
Swietenia macrophylla	*caoba*	*punab, ka'wakche´*
Malmea depressa	*elemuy*	*box elemuy*
*Platymiscium yucatanensis**	*granadillo*	*subin che´*
*Krugiodendron ferreum**	*quiebra hacha*	*chintok*
Dendropanax arboreus	*sac chaká*	*sak chakah*
*Cedrella odorata o mexicana**	*cedro*	*kú´che´*
Vitex gaumeri	*barrabás*	*ya´ax niik*
*Sweetia panamensis**	*chakte vigas*	*chakte'*
Sickingia salvadorensis	*chacte cok*	*chakte´kok, chactemuch*
Cordia dodecandra	*ciricote, siricote*	*kopté*
Pimenta dioica	*pimienta*	*ixnabakuk*
*Protium copal**	*copal, trementino*	*pom*
*Metopium brownei**	*chechem negro*	*box chechem*
Guettarda combsii	*papelillo*	*tostab*
Pouteria campechiana	*canistel*	*kániste´*
Thouinia paucidentata	*kamchunup, hueso de tigre*	*kanchup*
*Nectandra sanguinea**	*laurel blanco, aguacatillo*	*ekulub*
*Hampea trilobata**	*majagua*	*jool*
Calophyllum brasiliense	*bari, santamaría*	?
*Ficus sp.**	*matapalo, amate*	?
*Piscidia piscipula **	*jabín*	*ja´abin*
*Bauhinia herrerae **	*pata de vaca*	*k'ibich*
*Amyris elemifera **	*palo de gas*	?
Stratum between 7 and 12 m high:		
*Gymnopodium antigonoides**	*tzitzilche*	*ts´iits´il ché*
Hampea trilobata	*majagua*	*jool*
Malmea depressa	*elemuy*	*box elemuy*
*Krugiodendron ferreum**	*quiebrahacha*	*chimtok*
*Nectandra sanguinea**	*laurel blanco, aguacatillo*	*ekulub*
Guettarda combsii	*papelillo*	*tostab*
*Brosimum alicastrum**	*ramón*	*chak óox*
Acacia gaumeri	*katzim*	*box katsim*
*Protium copal**	*copal*	*pom*
*Bursera simaruba**	*palo mulato*	*chakah*
Stratum between 1 and 4 meters:		
*Piper sp.**	*cordoncillo*	?
*Nectandra sanguinea**	*laurel blanco, aguacatillo*	*ekulub*
*Chamaedorea sp.**	*xate hembra*	*xate*
Eugenia sp.	?	*kankabul*
Cryosophila argentea	*escobal*	*kum*
Sabal sp.*	*huano de sombrero*	*xa´an*

Table 4.1. Calakmul medium semi-evergreen species in each stratum. The asterisks indicate the species that were observed in the field (INEGI, 2002; Lundell, 1937 and my observations).

Figure 4.3. Calakmul medium semi-evergreen forest. Each white line divides the four vegetation strata.

Figure 4.4. Low semi-evergreen forest lower strata. In the distance can be observed a white tree trunk that forms part of the upper strata. Some areas of the akalche are impenetrable since fallen trees and lianas form complex meshes.

Figure 4.5. Vegetation of Calakmul largest aguada. This same area is covered by water during the rainy season. During the dry period, many animals come here to drink and eat.

The species of each of the strata are mentioned in Table 4.2. This listing was obtained from INEGI sample point 89 (INEGI, 2002), from Lundell (Lundell, 1937) and my observations in the field. The asterisks indicate the species observed in the field. Some areas of the *akalches* are practically impenetrable. The soils of these areas have high levels of organic matter and much clay.

Water reservoirs have distinctive vegetation reduced to an association of saw-toothed sedges and low, shrubby species sub-aquatic and aquatic plants (Lundell, 1937; See Figure 4.5). Unfortunately, the species that grow here are not very well studied (INEGI, 2002 did not register this kind of vegetation in peripheral sample points) but some of them floating water plants as *Pistia stratiotes* (*lechuga de agua;* water lettuce), *Eichhornia crassipes* (*jacinto de agua;* water hyacinth), *Nymphaea ampla* (*flor de agua;* water lily) and *Salvinia auriculata* (*oreja de agua;* eared watermoss) have been reported in the area (SEMARNAP, 1995).

After recognizing the three main types of vegetation a map was generated. With the INEGI ortho-photograph scale 1:20 000 as the base image (INEGI, 2001b), each type was delineated using Autocad Map 2002 software. Then a vectorial map was generated and each type of vegetation was quantified. The results are shown in Figure 4.6.

The vegetation types were interpolated taking in count an area of 64.605 km^2. Although all this area was not explored (particularly some sections of *Bajo* El Laberinto), the observation of the texture and characteristics in INEGI aerial photographs (INEGI, 2001a), ortho-photographs (INEGI, 2001b), Aster and Landsat images was useful for making these assumptions. The interpolated area includes the Calakmul promontory as well as the upper portion of *Bajo* El Laberinto. The results are shown in the Table 4.3 and Figure 4.7.

Almost two thirds of the area is represented by *akalche*, in particular by the upper portion of *Bajo* el Laberinto and its northern branches. In contrast, the aquatic vegetation that surrounds reservoirs is poorly represented (0.14%). Finally, only one third of the vegetation is medium semi-evergreen forest.

Studies of Past Vegetation

One of the most important objectives of this dissertation is to elucidate a panorama of the vegetation of Calakmul and its periphery when Maya inhabited the site. As Lundell points out regarding the past vegetation (Lundell, 1937:10), probably the greater part of the upland forest of a site was felled for agricultural purposes. Hence, what we may find in a current Maya rainforest is the anthropogenic result of previous managed vegetation.

When the research proposal of this study was conceived, it was contemplated to study the fossil pollen from the *aguadas* and from El Laberinto *bajo* of Calakmul in order to approach the vegetation succession and in order to detect primordial species. Two samples from two *aguadas*, the biggest one and the small one located next to structure VII, and two from the *akalche* in the 2003 season were obtained with special sampling corers. The samples were immediately taken to Dr. Gerald Islebe of Ecosur, Chetumal. After his team made preliminary analyses, they concluded that there was not enough preserved pollen for performing a complete study. Although this was a disappointing result, since many resources and much time were invested in this study, this

Scientific name	Spanish name	Maya name
Upper stratum, 10 m high:		
*Haematoxylon campechianum**	*palo tinto, palo de Campeche*	*ek'*
*Metopium brownei**	*chechem negro*	*box chechem*
Gymnopodium antigonoides	*tzitzilche*	*ts'iits'il ché*
*Manilkara zapota**	*chicle, zapote, chicozapote*	*ya'*
Erythroxylon areolatum	*cordiache*	*?*
Bucida buceras	*almendra de río*	*pukté*
Cordia dodecandra	*ciricote, siricote*	*kopté*
Hampea trilobata	*mahagua*	*jool*
*Cameraria latifolia**	*chechem blanco*	*sak chechem*
Croton sp	?	*?*
Eugenia sp.	?	*kankanbul*
Pithecellobium sp.	?	*?*
Vitex gaumeri	*carrete*	*yax'nik*
Guettarda combsii	*manzanillo*	*tastab*
Randia aculeata	?	*ak' ank'ax, peech kitam*
Ficus sp. *	*matapalo, ama te*	*?*
*Bursera simaruba**	*palo mulato*	*chakah*
*Brosimum alicastrum**	*ramón*	*chak óox*
*Bauhinia herrerae **	*pata de vaca*	*k'ibich*
Lower stratum, 2 meters high:		
*Cameraria latifolia**	*chechem blanco*	*sak chechem*
*Manilkara zapota**	*chicozapote, chicle*	*ya'*
Croton reflexifolius	*palito de sabana*	*pereskutz*
Randia sp	?	*e.g. ak' ank'ax*
Jacquinia aurantiaca	*sicate*	*chak sik*

Table 4.2. Calakmul low semi-evergreen species in each stratum. The asterisks indicate the species that were observed in the field (INEGI, 2002; Lundell, 1937 and my observations).

Vegetation Type	Extension m^2	Total Area %
Medium semi-evergreen forest	23 273 297.04	36.02
Low semi-evergreen forest	41 237 536.30	63.83
Aquatic vegetation	94 943.23	0.15
Vegetation Total	64 605 776.57	100.00

Table 4.3. Extension of Calakmul vegetation types and percentage of total area.

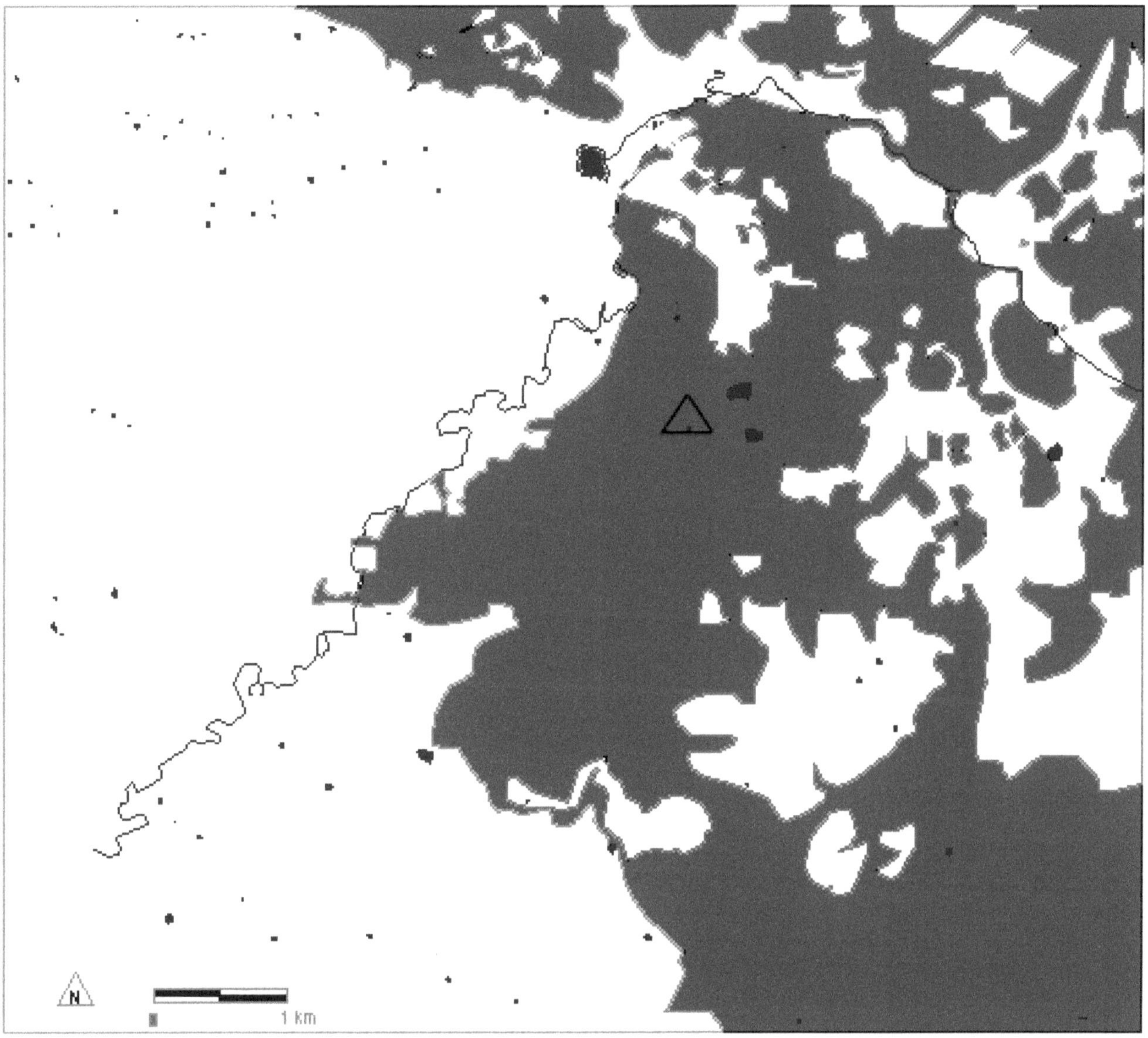

Figure 4.6. Vegetation types of Calakmul and periphery. Information extracted from INEGI ortho-photograph (INEGI, 2001b), scale 1:20 000, as well as field reconnaissance. Low semi-evergreen forest is represented in white; medium semi-evergreen forest in dark gray. Water reservoir vegetation is drawn in black. The civic-ceremonial center is indicated by the black triangle.

experience can be useful for future researchers in order to avoid sampling in these same places

As part of the research related with Tomb 4 in Substructure IIB, pollen samples were submitted for analysis. As a result Mc Clung, Tavera and Casales affirm that they found some Amaranthus and Chenopodium that could have been used for human consumption. Potential medicinal plants found were Ambrosia, Nuphar, Myrca, Alnus, Rubieaceas, Mimosas, Bigoniaceas and Chenopodium (Carrasco et al., 1999a). They also identified pollen pertaining to genera as Pinus, Alnus, Ulmos, Acer, Cupressus and Eucaliphtus that do not belong to the Calakmul area. Hence, there is a possibility that the pollen of these species somehow reached from outside or that the samples were contaminated.

In other studies of Tomb 4, after analyzing the macro and micro botanical fragments, they found a fiber "that is not cotton" (Carrasco et al., 1999a). Similarly, restorers reported that in this same tomb *Bursera simaruba* (*chakah*) wood, and a palm structure that was used to support the headdress of the personage were found (Carrasco et al., 2000).

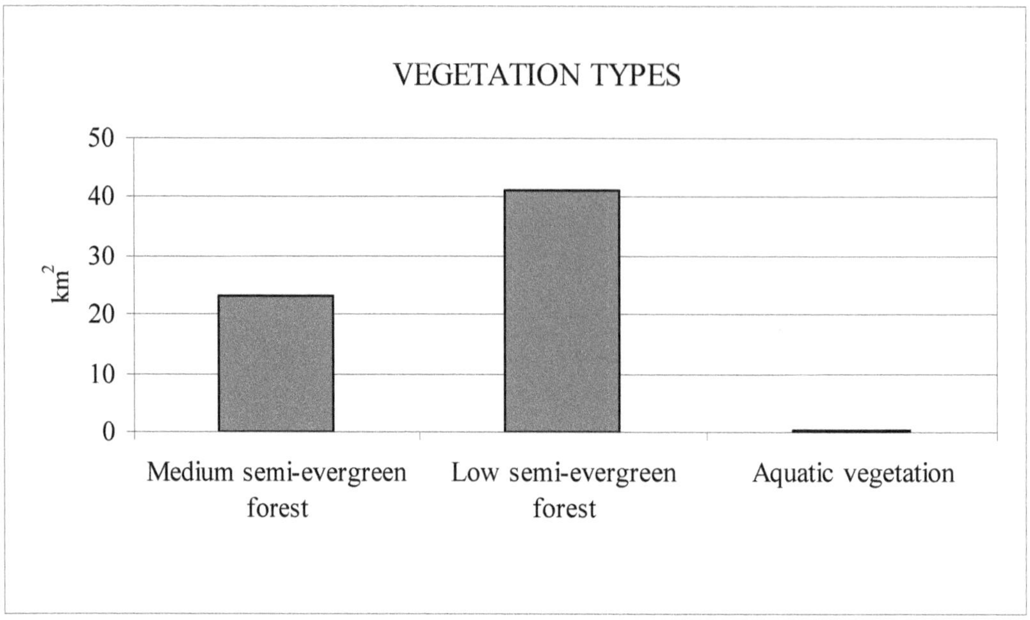

Figure 4.7. Extension of Calakmul and its periphery vegetation types.

Research in other sites relatively near Calakmul provides additional vegetation data. Gunn and colleagues recognized large grass cell types and canopy vegetation through the identification of phytoliths in sediment cores extracted from the *bajos* El Laberinto and El Ramonal (Gunn et al., 2002). Unfortunately, they did not indicate the precise location of their test pits (they affirm that they were sampled in the vicinity of Villahermosa and El Ramonal archaeological site). Since both sample sites are approximately 20 kilometers far away from Calakmul site center, these results only serve to suggest the possibility of similar vegetation at Calakmul. I had the opportunity to observe Naachtun vegetation characteristics during the 2004 season. Naachtun, an ancient Maya site, is located 30 kilometers to the south of Calakmul and its vegetation differs noticeably from that of this site. Hence, the interpolation of the vegetation should be taken with that in mind.

After examining the present day vegetation of Calakmul some important points stand out. The first one is that today's rainforest is the result of an ancient anthropogenic management (Barrera et al., 1977). This means that some species were directly or indirectly selected and that the rainforest is not a pristine setting. As Lundell points out, trees such as the *zapote* and *ramón* were doubtlessly protected and planted by the Maya, and hence gained an advantage over the other vegetation when the uplands were abandoned (Lundell, 1937). A recent study of Calakmul vegetation confirms this hypothesis (Brown, 2005). Moreover, at field one can detect that these same species are very frequent in Calakmul as well as *guayas* and *chakah.*

The second point is that the ancient Maya people might have had the *milpa* system over a large part of the area, even in the *bajos*. Regarding this matter, Lundell affirms that continued *milpa* agriculture and fire destruction have so changed conditions in some localities that savannas exist in some Peten areas, even though not particularly favored by the climate and soils (Lundell, 1937). The large quantity of human vestiges makes me think that Maya would have needed to manage all the available land in order to satisfy their primary needs.

Finally, the rainforest that we observe today has been modified by *chicle* harvesting and timber extraction. As a consequence, precious timber trees have disappeared and other plants have taken their place. Although Lundell reports mahogany in the Peten area I did not see a single tree of this species in the field.

After observing the high density of Maya habitation in the site of Calakmul and in the periphery of *Bajo* El Laberinto, it is possible that few areas were undisturbed, at least during the Late Classic. During the 2003 and 2004 field reconnaissance, almost continuous, ancient house-mounds were detected. Those vestiges indicate an elevated population number and less space for undisturbed vegetation areas. Impoverished forests limited timber, medicinal, apicultural and plant resources, a situation that probably had dramatic consequences during the Late Preclassic in Calakmul.

CHAPTER 5. SOILS AND LAND EVALUATION OF CALAKMUL

"...so that everything that there [Yucatan] is and that grows in it grows better and more abundantly amongst the rocks than in earth, because on the earth which happens to be in some parts, neither do trees grow nor are there any, nor the Indians sow their seeds, nor is there anything except grass. And among the stones and over them they sow and all their seeds spring up and all the trees grow and some so large and beautiful that are marvelous to see. The cause of this is, I believe, that there is more moisture and it is preserved more in the rocks than in the earth."

Fray Diego de Landa in Tozzer, 1941:186.

The Importance of Soils

Soil constitutes the matrix in which organisms live. It is interesting to know that the Earth's area useful for agriculture is approximately 7.5% of its surface. With only this extension of land, food is procured for all humans (The Natural Resources Conservation Service, Syracuse. Department of Agriculture, http://soil.gsfc.nasa.gov/app_soil/hmsoil.htm., accessed July 2004). Hence it is very important to know soil characteristics in order evaluate their potential role in agriculture or construction. Determining their agricultural potential, urban suitability or land and water interaction can help to evaluate the strategies used by societies for subsisting during a specific period. Furthermore, knowing the processes that affect soils can prevent harmful practices that may cause soil malnutrition or desertification.

The study of soils is intrinsic for archaeological research. A soil profile gives a picture of geological, climatic or biological conditions as well as human history. Studying soils give us information related with the site's stratigraphy, the periods of inhabitation or the intensity of the occupation of a place. Soils can be used for reconstructing past landscapes and landscape evolution, estimating the age of surfaces and depositional episodes, and for providing physical and chemical indicators of human occupation (French, 1994). In terms of past environment, the soils are valuable for envisioning also settlement patterns, clearance, erosion, cultivation, agricultural activities and use of space.

What is a Soil?

The soil is composed of a mix of mineral particles, organic matter, microorganisms and disintegrated rocks (Eyhorn et al., 2002). Each soil is the result of the interaction of parent material, relief, climate and living organisms during a certain time span (Jenny, 1941). Soils are dynamic entities, continually changing.

Many definitions can be found in relation with soils. The United States of America Soil Survey Staff Division has been for many years a recognized authority for classifying soils and developing continually new methods for edaphologists. They provide the following definition of soils:

"the collection of natural bodies in the Earth's surface, in places modified or even made by man of earthy materials, containing living matter and supporting or capable of supporting plants out-of-doors. Its upper limit is air or shallow water. At its margins it grades to deep water or to barren areas of rock or ice. Its lower limit to the not-soil beneath is perhaps the most difficult to define. Soil includes the horizons near the surface that differ from the underlying rock material as a result of interactions, through time, of climate, living organisms, parent materials, and relief. In the few places where it contains thin-cemented horizons that are impermeable to roots, soil is as deep as the deepest horizon. More commonly soil grades at its lower margin to hard rock or to earthy materials virtually devoid of roots, animals, or marks of other biologic activity. The lower limit of soil, therefore, is normally the lower limit of biologic activity, which generally coincides with the common rooting depth of native perennial plants" (Soil Survey Staff, 1975 in USDA SSDV, 1993:5).

Although rather lengthy definition may look complicated, it eliminates many doubts of what and what is not a soil. Also this paragraph introduces the concept of 'horizon' or each one of the visible layers of a soil profile. Each horizon implies a series of processes that have affected the soil and hence present specific characteristics. For example, C horizon represents non-consolidated material present in the contact zone between soil and rock, or the R horizon represents the unweathered bedrock that is beneath all layers. The horizons of this study were identified according to FAO method for classification (León, 2003:159-169).

Soil Research in Yucatan Peninsula

As part of a global project for the identification of the soils of the world, FAO published a 1: 25 000 000 scale map. This World Soil Resources Map illustrates that the dominant soils in the east and a southwest coast portion of the Yucatan Peninsula are Histosols and Fluvisols, and Leptosols, Regosols and Gleysols in the rest of it (FAO, 2003). In the same way, the FAO-UNESCO Soil Map of the World, scale 1:5 000 000 (FAO-UNESCO, 1992) identify Rendzinas and Vertisols as the main soil units in the south central Yucatan Peninsula, where Calakmul Biosphere Reserve is located. Additionally, the Instituto Nacional de Geografía y Estadística (INEGI) digital map of soils, scale 1:20 000 000 shows that the soils of Calakmul area are Gleysols in the plains and Rendzinas in the hills (http://mapserver.inegi.gob.mx/map/datos_basicos/edafologia/?c=527; accessed June 20th, 2005).

Studies that describe the Yucatan Peninsula soils employing a more detailed scale have been done. Duch described Lithosols, Rendzinas, Vertisols and Gleysols in the Peninsula (Duch, 1989). In the same way, Francisco Bautista Zúñiga is gathering a soil database of the

Yucatan Peninsula edaphology with the objective of preparing a complete soil identification program using the last technological advances in software and satellite technology (personal communication, 2005). As part of this project, he has re-classified the soils of the Peninsula at a 1:250 000 scale (Bautista et al., 2005).

In the Maya area, archaeological research related with soil studies has been done. Studies in Eastern Puuc region (Dunning, 1992), western Puuc area (Isendahl, 2002), northern Yucatan Peninsula (Fedick and Morrison, 2004), Belize (Dunning et al., 1999; Muhs et al., 1985; Murtha, 2002), Bonampak (Aliphat et al., 1997; Geovannini, 1998), the Usumacinta River region (Aliphat, 1994) and Copan (Wingard, 1992; 1996) are just some examples of investigation that uses soils as a main object for the cultural interpretation of archaeological remains.

It is very important to state that actual Maya people classify soils (as do Western soil taxonomists) in a way that reflects their agricultural value. In northern Yucatan Peninsula different interviews have exposed the soil categories of Maya native people that have inherited this knowledge from their ascendants (e.g. Bautista et al., 2006). Table 5.1 shows the equivalences between Yucatan Peninsula Maya and FAO soil classifications, as well as Yukatec Maya dictionary definitions of some of those soils. Although this is a present-day classification, these categories are a good reference for emphasizing the soil characteristics of the area and their value for cultivation. The Maya soil classification has main features for identifying soils such as color and quantity of stones in contrast with FAO taxonomy, which requires more environmental variables for the same purpose.

In this same table, it is also evident that each study denominates the majority of the soil folk Maya categories differently (Bautista et al., 2006; Isendhal, 2002 and UQROO-PEOT, 2000). They only coincide with Lithic Leptosols and Rendzic Leptosols. Perhaps, the fact that those studies were done in different places, with different methodology is the reason for those unrelated interpretations.

Unfortunately, there are not long-term inhabitants to interrogate near the site of Calakmul in regard to this topic. Besides, many of the Calakmul archaeological project workers were born in other states of the Mexican Republic and therefore are not the best source for providing this kind of information.

Soil Research in Calakmul Area

The report prepared by FAO groups the soils of the Calakmul Reserve as part of the Xpuhil Series Association (FAO, 1970). Years passed before other researchers investigated the area of Calakmul itself. After a good road was opened in the 1980s, the studies related to the description of the environment proliferated. The interest in the *bajos* and how could they support large populations was a major question in archaeological projects in Belize, Guatemala, Honduras and Mexico. One of these projects was undertaken near two archaeological sites, El Laberinto and El Ramonal, both within a distance of about 20 km from Calakmul archaeological site. Five soil cores were excavated in El Laberinto and El Ramonal *bajos* (Gunn et al., 2002). The team identified Vertisols in the center of the *bajos* with elevated proportions of gypsum and high salinity. In contrast, the edges of the *bajos* were less saline, with high levels of calcium. Gunn and colleagues proposed that the high levels of salinity in the lower strata of their excavation units indicated that *bajos* were relatively saline under pre-urban conditions, at least seasonally. They also identified fossil phytoliths and observed large grass cell types and canopy vegetation.

Closer to Calakmul, biologists Morales and Magaña identified Lithosols, Rendzinas, Vertisols, Gleysols and Fluvisols soils in the Biosphere Reserve (Morales and Magaña, 2001). In the archaeological zone, Morales identified associated Vertisols and Rendzinas, and Gleysols with Vertisols near the *aguadas* (Morales, 1999). She affirms that these soils correspond with the ones that Jacob described for Nakbe in Guatemala (Jacob, 1995). In addition, Naraya Carrasco (Carrasco et al., 2001) completed six soil test pits toward the north of the Great Acropolis with the aim of describing the soils that surrounded the site core. As a result she found Rendzinas, Vertic Gleysols and Lithosols. Although the above-mentioned studies identified the main soils of the area, they did not describe their characteristics and distribution in the site.

The Purpose of the Soil Survey in the Context of this Study

One of the principal aims of this study is to carry out a soil survey in order to understand better how ancient Maya related to their environment. The objective of a soil survey is "to provide geographic information on soils, which can be used to correlate with vegetation data to obtain a more complete picture of a given ecosystem". And following the same line, another important purpose is "to provide the basis for developing an ecologically sound land use" (Breimer et al., 1986:15). As Dent and Young state:

Soil (Maya)	Isendhal , 2002:69	Bautista et al., 2006 WRB classification	UQROO-PEOT, 2000	Barrera Vázquez, 1995 Dictionary	Bastarrechea et al., 1992 Dictionary
Tsek'el	Lithic Leptosol	Lithic Leptosol	Lithic Leptosol	earth full of stones and not good to sow	very stony soil
Kaccab	Rendzic Leptosol	----	----	earth good to sow	----
Chac lu'um	Rhodic Leptosol	Endoleptic Cambisol and Epileptic Cambisol	Chromic Cambisol	----	----
Pus lu'um	Rendzic Leptosol	Rendzic Leptosol	Rendzic Leptosol	dry earth without stones	----
Ek lu'um	Vertic Rendzic Leptosol	Luvisol	----	earth good for bread	----
Box lu'um	----	Lithic Leptosol, Hiperskeletal Leptosol and Calcisol	----	----	----
Chich lu'um	----	Leptosol and Calcisol	----	----	----
Hay lu'um	----	Lithic Leptosol	----	----	----
Yax Hoom	Chromic Vertisol	Lixisol	Eutric and Distric Vertisol	----	----
K'ankab-Tsek'el	Lithic Rhodic Nitosol	----	----	red or red-brown earth	red soil with humus and carbonates
K'ankab- Ak'alche	Chromic Luvisol	----	----	----	----
Ak'alche	----	Vertisol	Mollic and Eutric Gleysol	----	lagoon or *ciénega* surrounded by trees
K'ankab	----	Cambisol and Luvisol	Chromic Luvisol	----	----
Huntunich	----	----	Calcaric Regosol	----	----
Chaltún	----	Lithic leptosol	----	----	----
Ch'och'ol	----	Hiperskeletal Leptosol	----	----	----

Table 5.1. Soil classification equivalence between FAO and Maya soil folk taxonomy as well as Yukatek Maya - Spanish dictionary definitions of those soils.

"the practical purpose of soil survey is to enable more numerous, more accurate and more useful predictions to be make for specific purposes than could have been made otherwise [i.e., in the absence of location-specific information about soils]. To achieve this purpose, it is necessary to determine the pattern of the soil cover; and to divide this pattern into relatively homogeneous units; to map the distribution of these units, so enabling the soil properties over any area to be predicted; and to characterize the mapped units in such a way that useful statements can be made about their land use potential and response to changes in management" (Dent and Young, 1981:1).

Taking into account the objectives of this study, the soil survey method that was followed, was the one proposed by Breimer and colleagues as part of the UNESCO intergovernmental Programme on Man and Biosphere (Breimer et al., 1986). This program aims to develop, within the natural and social sciences, a basis for the rational use and conservation of the resources of the biosphere and the improvement of the relationship between man and the environment. In particular these guidelines have been applied in developing countries such as Kenya, Mexico, Gabon and Indonesia.

The definition of soil survey that has been used in the context of this study is the following one:

"A soil survey describes the characteristics of the soils in a given area, classifies the soils according to a standard system of classification, plots the boundaries of the soils on a map, and makes predictions about the behavior of soils. The different uses of the soils and how the response of management affects them are considered. The information collected in a soil survey helps in the development of land-use plans and evaluates and predicts the effects of land use on the environment." (USDA Soil Survey Division Staff, 1993:1).

Considering the available maps and images of Calakmul Biosphere Reserve and the project schedule and objectives, a semi-detailed soil survey was chosen. A semi-detailed soil survey involves maps with scales of 1:25 000 to 1:100 000 in an area that normally ranges from 50 to 1000 km^2. Some important characteristics of these surveys are as follows: soil associations can be mapped in heterogeneous areas, local variation sometimes can be ignored, and aerial photographs are crucial for delineating soil boundaries which can then be checked in the field (Breimer et al., 1986:16).

The first step for soil surveying was the observation of the Calakmul landforms and its parent material. This approach has been called "physiognomic-lithomorphic" (Sombroek and Van de Weg, 1980). First, landforms are described in the soil map legend considering physiography and altitude, for example, uplands (U) or dunes (D). Each landform should have similar physical characteristics and hence a uniform origin. Second, the basic geological information is added to the landform legend, as for example MF, mountains (M) with rocks rich in (F) ferromagnesian minerals (Breimer et al., 1986). The landforms and geology of Calakmul have already been described in Chapter 3 and are illustrated in Figure 3.6.

Finally, the main soil unit, soil association and soil complex are described. The descriptions refer to the solum, usually A and B horizons to 100 cm, unless the bedrock occurs at a shallower depth (Breimer et al., 1986). The described characteristics are: drainage conditions, effective soil depth, color, mottling, consistence, calcareousness, salinity, sodicity, stoniness, rockiness, cracking, texture and observations (USDA SSDS, 1993). An example would be MFr: Mountains (M), gneisses rich in ferromagnesian minerals (F) well drained, predominantly deep, dark reddish brown to yellowish red, friable, sandy clay loam to clay loam with moderately thick, low humic topsoil, in places stony and rocky. In addition, particular observations and the general slope class can be added. Lastly, each mapping unit description should be followed by a taxonomic classification (in this case the FAO system [FAO, 1974]), which is valuable for data extrapolation (Breimer et al., 1986).

For this study, the considered taxonomic classification was the one proposed by FAO for a number of reasons (FAO, 1974). First, INEGI, the Mexican government agency that has mapped the entire Republic, uses this classification, so the results are comparable with the maps of other areas. Second, the identification of soils is based in measurable physical and chemical criteria that can be observed in each profile and that usually do not consider the climate as a criterion for classification (Breimer et al., 1986). Third, the FAO classification is definitively much more understandable for non-edaphologists than the United States soil taxonomy.

Procedures for Soil Survey

Maps and images of different scales were used for this investigation. The topographic maps 1:50 000 (INEGI, 1987), 1:20 000 (SIGSA, 2004) and 1:6250 (May Hau et al., 1990) of the area were digitized in order to obtain a base map for performing the subsequent analysis. Topographic maps that previously were generated by other researchers were considered too (García Gil, 2002; Lundell, 1933; Ruppert y Denison, 1943).

Additional INEGI small-scale (1:250 000) maps were also of great value. Maps of vegetation and soil use (INEGI, 1985; SPP, 1981), superficial and subterraneous hydrology (INEGI, 1988; 1989), geology (INEGI, 1999) and climate (SPP, 1980) were revised. Other helpful charts were the Mexican hydrological regions map (Díaz Ponce, 1999) and the FAO soil map (FAO, 2003).

Aerial photographs and other remote sensing images were also used. These pictures have multiple uses in archaeology. The objective in using them is to recognize relevant elements from the relief and vegetation of the terrain that sometimes even in field cannot easily be seen. Aerial photographs scale 1:75,000 (INEGI, 2001a), ortho-photographs (INEGI, 2001b), a multi-band Aster image of Campeche, as well as a Landsat, ESRI Earthsat (from year 2004) and Skyline Software Systems Terra (from year 2003) images of Calakmul were taken into account for image analyses.

The six group of elements that were recognized in aerial photographs in order to distinguish soils differences (as well as relevant landscape features) in conjunction with field studies, and presented in decreasing order of correlation were the following ones (Breimer et al., 1986:33; Vink, 1968):

1. Elements related with a positive, direct correlation with an aspect or aspects of the soil itself (for example, waterlogged soils).
2. Elements related to the general morphology of the terrain (such as relief, river patterns or individual rivers and streams).
3. Elements related to special aspects of the terrain (such as erosion).
4. Elements related to the vegetation cover (such as trees species or land use).
5. Elements related to human activities (as for example canals, ditches, roads and buildings).
6. Inferred elements or elements based on "converging evidence" (such as drainage conditions, parent material and micro-relief).

During and after the field seasons, and with the use of a global positioning system (GPS) for locating each point, the soil boundaries were drawn up by hand in aerial photographs. It was a relatively easy task, since scales were similar and there were many landscape features that could be readily identified. The final results are based on these photographs (INEGI, 2001b).

The identification of the soils of the area and their combinations was achieved through the field survey of the Calakmul region. The survey was done by adopting the transect method, in particular, following the edaphological '*catena*' procedure. A *catena* is a soil sequence that has similar parent material, with equivalent age in related climatic areas, but which characteristics differ because of drainage and relief variation (Tamhane et al., 1979). The number of observations depends on the complexity of the landscape and the final scale of the map (Sombroek and Van de Weg, 1980). Following the transect approach, a limited number of representative transects that cross different landforms as much a possible were selected. Transect reconnaissance was done then, by identifying zones with the most differences. These differences are given by the distinct geological and physiographic elements. Five *catenas* were sampled in order to cover the terrain differences (See Figure 5.1). In each sample unit a pit of 1x1 meters was excavated, its profile described and a sample taken from each soil horizon. The forms used for description are presented in Appendix 2.

Transects were planned both perpendicular to and parallel to the karstic spine in order to identify major differences, with an emphasis on the promontory, and around the permanent reservoirs. Twenty-seven pits were done and only fourteen representative profiles were analyzed (shown with capital letters in Figure 5.1). As part of the reconnaissance, thirteen additional pits were made with an auger in order to examine soil characteristics, and if they were already observed, they were not described further. This decision was done for practical reasons since any investigation has time and limited resources. Selected transects did not coincide with roads because usually these are done in elevated terrain. The explored area can be appreciated in Figure 5.1. For this study, the number of pits was enough for the final scale and objectives of the project. After sampling, the soil was dried for several days and prepared for performing the required analyses developed by different soil scientists.

The analyses of the soil horizons were done in order to describe the physical and chemical characteristics, essential for classification. In the same way the information was useful for defining the soil boundaries of the study area. In addition, data related with soil capability was generated in order to propose a map of potentially suitable agricultural land. Although these analyses have a standardized tendency, it is possible to find several versions of the same experiment, and many manuals with different protocols. As a member of previous ecological projects, I had the opportunity to perform different protocols for soil analysis, and the most practical and efficient ones were chosen. In other words, I adapted the best protocol according to the objectives of this project.

The physical and chemical soil analyses performed for this study were: structure (León, 2003), texture (La Motte, n. d.), organic matter content (Nuñez, 1977), pH (Porta et al., 1994), conductivity (Aguilar, 1987; IRENAT, n. d.), color (Munsell Color Company Inc., 1954), apparent density (ICUAP, n. d.) and phosphates (Hanna Mehlich Extraction Procedure and Phosphorus Test for Soil Fact Sheet; Mallarino and Sawyer, 1999). All the information was recorded in the respective field and laboratory diaries. The protocol is available in Appendix 1.

The properties of the soil have been divided into physical, chemical and biological. A table based in the most important soil properties leading to its classification and evaluation is included in Table 5.2. These properties were evaluated as part of this study. The complete final results can be consulted in Appendix 3.

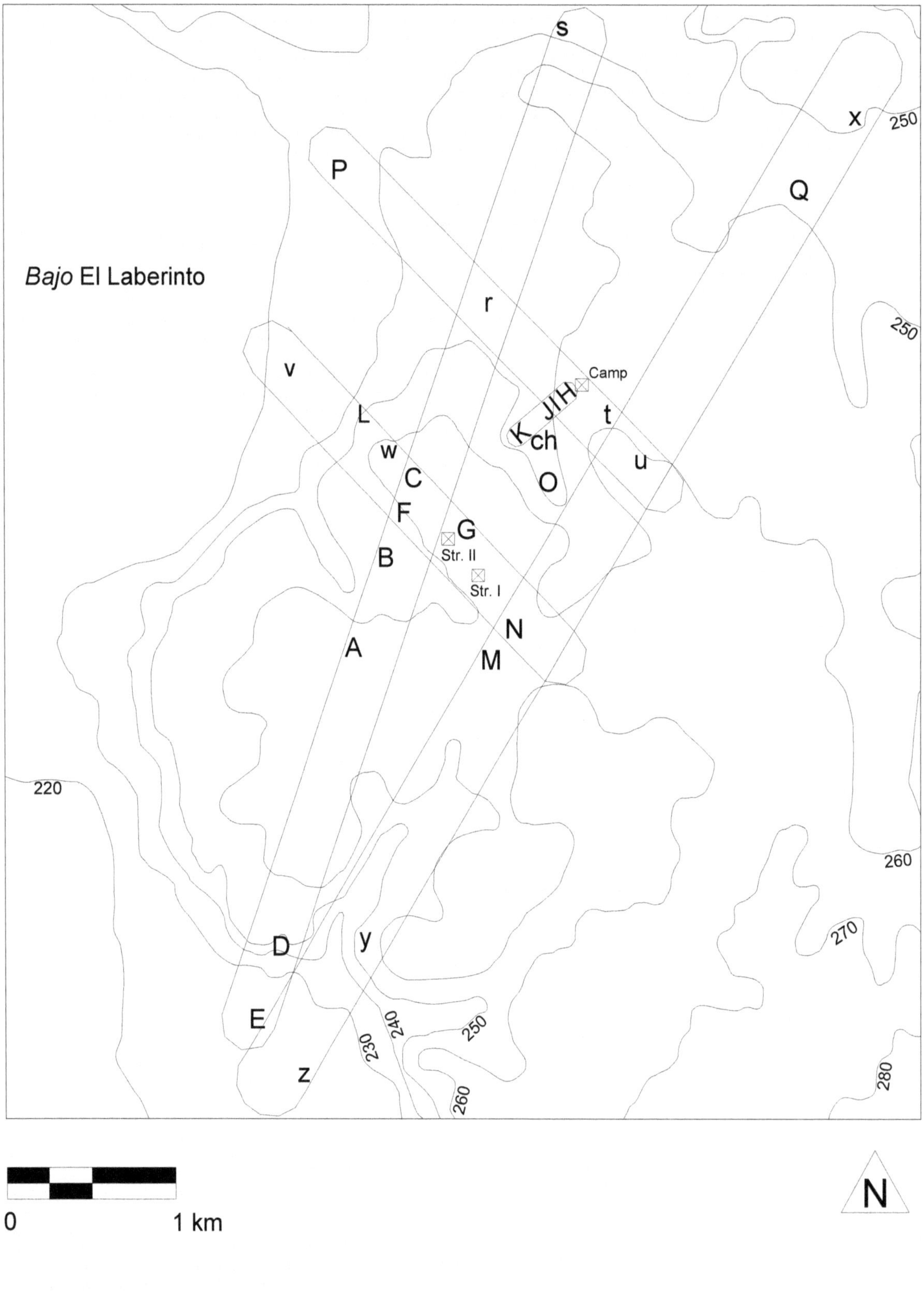

Figure 5.1. Location of the edaphological catenas. Each letter indicates a pit. Str. means structure.

Soil Properties	Features
Climate	Pluvial precipitation Temperature Winds
Vegetal Cover	Vegetation
Surface Properties	Physiographic position Slope Micro-relief Erosion Superficial drainage
Physical Conditions	Color Effective depth Density Porosity Permeability Texture Stoniness Structure Humidity
Chemical Properties	Organic matter pH Calcium carbonates Salts Phosphates Fertility
Land Productivity	Result of the mentioned properties

Table 5.2. Soil properties classification (based in Storie, 1970).

Surface Properties

- Physiographic Position

The physical features of the terrain are the beginning point for landscape studies. The form of the relief, its composition, genesis and arrangement should be described as best as possible. The description of Calakmul physiographic units has been presented in Chapter 3.

- Slope

Terrain inclination is a very important factor for soil formation and erosion issues. Farming practices are adapted to terrain slope. The slope length is as significant as its decline percentage. For example, large slopes allow water accumulation if ditches are built. The slope classification used principally in Mexico is shown in Table 5.3.

In an area of 64.60 km^2, the map of slopes was generated using ArcGIS 3D Analyst Version 8.3. After observing the histogram, nine classes were generated with the Jenks method of natural breaks (Figure 5.2). As it is can be observed, the great majority of the area is flat and almost flat. Even in higher places such as the Calakmul promontory and the eastern hills, micro-valleys are formed. Higher inclinations, between 5 to 40%, are seen in promontory boundaries and around the plateau hills.

Category	Degree	Percent
Flat	0-0.9	0-1.55
Almost flat	1-1.9	1.55-3.49
Slightly inclined	2-5	3.49-8.7
Moderately inclined	5-12	8.7-21.25
Inclined	12-20	21.25-36.39
Very inclined	20-40	36.39-83.90
Steep	40-60	83.90-173.20
Mountainous	>60	173.20

Table 5.3. Slope percent classification (León, 2003).

- Micro-relief

Careful consideration of slight topographical variations in the terrain, such as minor depressions or mounds, is important since they naturally modify the flow or drainage of water, the ways in which vegetation stands or the way the wind acts on the surface of the land. The area of the El Laberinto has a relatively homogeneous relief with a slight slope toward its center. In contrast, the low-lying *bajo* area situated on the north and east of the archaeological zone is interspersed with elevated areas, sometimes forming small islands that resemble permanent *petenes*. The topography of the hilly region near the site center is heterogeneous, presenting no discernable pattern of relief. Hills and mounds of different sizes, as well as depressions and micro-valleys are observed throughout this area.

- Erosion

Erosion is the displacement of the soil by agents such as wind, water, gravity, ice and living organisms, including humans. Erosion has existed for thousands of years as a natural process but humans have increased the intensity and observable effects of this phenomenon in the last millennia. The degree of soil erosion is measured by observing the surface layer of a specific area (León, 2003). It can be classified in three general grades: slight.

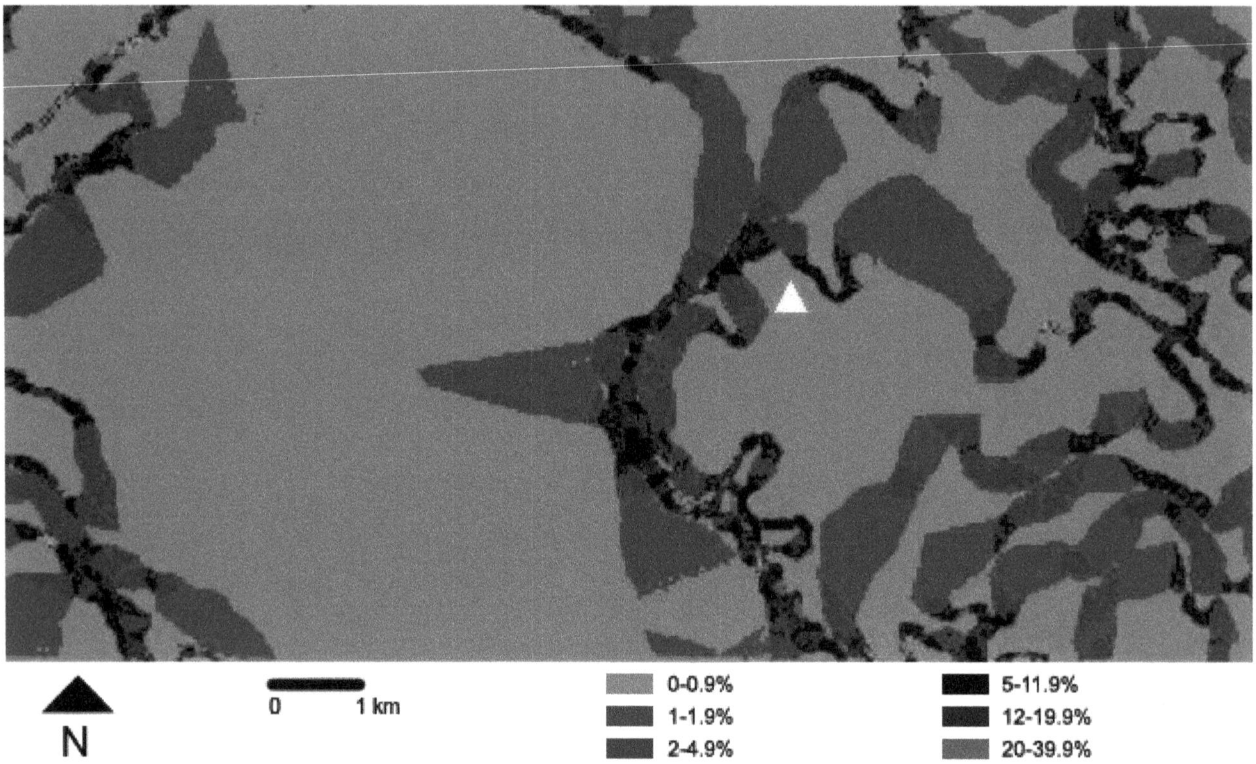

Figure 5.2. Slope map of Calakmul. The site's civic-ceremonial center is indicated by the white triangle.

erosion (loss of < 25%), moderate erosion (loss of 25 to 75%) and severe erosion (loss of >75%). Moderate erosion was observed in the context of the monumental area of the site. Diluted clays were noted in the Great Acropolis as a consequence of rain and the absence of proper drainage within this place. Additionally, the ravines located to the south and west of the site promontory increase terrain erosion in the rainy season. However, since a vegetation cover is present, erosion is not a major problem in the area.

- Superficial Drainage

Superficial drainage is the quantity of water that drains through the soil (León, 2003). It can been classified as good if water does not accumulates on the surface, moderate if it drains slowly and/or accumulates for short periods, and poor if water accrues and stagnates for long periods. Poor drainage is observed in *akalche* areas. When it rains water can be accumulated for several days. In contrast, good drainage is observed in hills and micro-valleys since the soil characteristics permit a rapid rate of infiltration and slopes channel water to the flat areas.

Physical Properties

The physical properties of the soil are valuable in terms of describing the interaction of its component elements. They are very important for agriculture since temperature, water, roots and living organisms will relate to each other according to them. The characteristics of these properties are described below; the sampled soils properties can be consulted in Appendix 3.

- Color

In order to register color, a well-known reference is the Munsell Soil Color Chart (Munsell Color Company Inc., 1954). Each color in this chart is classified according to the hue or dominant spectral color, the value or the relative darkness of the color and the chroma or the purity of the color. The color reflects soil properties. Its appearance varies according to lighting conditions or humidity, so in order to give a consistent description, it should be moistened. Calakmul soils have basically black, brownish and grayish colors, being white the limestone parent material. Soils with grayish color reflect water-logged conditions.

- Effective Depth

The effective depth is the available space in which the roots of the plants will be able to penetrate without interference for the purpose of obtaining water and nutrients (León, 2003). Penetration of roots will be limited by rock, gravel, water (phreatic mantle) and parent material. According to their effective depth, soils are classified as deep soils (>1 m), moderately deep soils (0.6 to 1 m), shallow (0.25 to 0.6 m) and thin (<0.25 m).

Roots are important in terms of preserving soil structure. They allow water permeation through pores, promote better drainage, improve soil pH, fix available nutrients and create a good place for earthworms. In order to grow, roots absorb minerals and nutrients in soil particles, aided by microorganisms and weathering. The soils of Calakmul have a variable effective depth, since they are alternately moderately deep, shallow, and thin.

- Density and Porosity

Apparent density is a measure of the degree of compaction of a soil and of how soil particles are clumped together. The size of the pores determines how the soil interacts with water. For example, large pores allow faster water filtration (Eyhorn et al., 2002). Porous soils will have low apparent density, in contrast to compact ones that will exhibit high apparent density. The apparent density of Calakmul soils varies from 1 to 1.03.

- Permeability

Permeability is the quality that affects the movement of water and air in the soil. It is very important in terms of determining and/or measuring degrees of root penetration, water absorption, internal drainage and nutrient infiltration. The permeability of each soil horizon within a specific profile is different. Permeability has been classified as slow, moderate, fast and very fast (León, 2003). In slow permeability soils, air cannot penetrate the soil, water is present for long periods, color is grayish, greenish or bluish, with specks, and great clay content is observed. Moderate permeability is a favorable condition for water and air movement. These soils are red or black, very suitable for agriculture. Fast permeability soils are those with sand in all layers. As a consequence, water passes through them rapidly, resulting in low fertility and aridity. Their subsoil is brilliant red. In the very fast permeability soils this condition is augmented. The permeability of the soils at Calakmul varies between fast and slow depending on their characteristics and location.

- Texture

The texture is the proportion of the soil particles according to their size (León, 2003). The particles, called separates, have been divided in coarse sand, sand, silt and clay. The separates recognized by the International Society of Soil Science are seen in Table 5.4.

Soils with fine texture have a high capacity for the retention of nutrients and water. They form a very difficult terrain for agriculture since clay becomes heavy with water. In contrast, soils with medium texture have a good balance of soil particles, are easy to work with, yet they still retain a fair degree of water and nutrients. These soils are known as loams. Silty loams and sandy clay loams retain water without water-logging and usually have the required nutrients. Sandy soils, with coarse texture are also easily manipulated, but they do not retain water and nutrients because of fast infiltration. As a consequence they are not very fertile. Too much sand or clay restricts vegetation and soil workability. In Calakmul the majority of the soil horizons have a fine texture.

Separates	Size in mm
Coarse sand	2.0 to 0.2
Fine sand	0.2 to 0.02
Silt	0.02 to 0.002
Clay	<0.002

Table 5.4. Soil separates size (Research Branch, 1976).

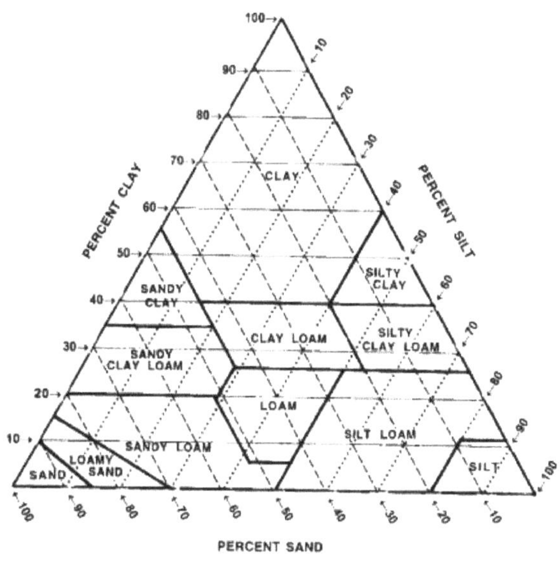

Figure 5.3. Soil texture triangle (USDA Soil Survey Division Staff, 1993).

The relative proportion of each soil separate determines its texture. With the aid of a textural triangle the relative percentage of each separate can be recognized. Each side of the triangle represents a separate, and at the point where the three converge the textural class is identified (See Figure 5.3). The soil textures observed in Calakmul are diverse. The textures found were clay, silty clay, silty

	Type		Kind	Class	Size (mm)
1	Structureless - no observable aggregation or no definite orderly arrangement around natural lines of weakness.	A.	Single grain - loose, incoherent mass of individual particles as in sands.		
		B.	Amorphous (massive) - a coherent mass showing no evidence of any distinct arrangement of soil particles.		
2	Blocklike-soil particles arranged around a point and bounded by flat or rounded surfaces.	A.	Blocky (angular blocky)- rectangular and flattened, vertices sharply angular	Fine blocky	<10
				Medium blocky	10-20
				Coarse blocky	20-50
				Very coarse blocky	>50
		B.	Subangular blocky-faces subrectangular, vertices mostly oblique, or subrounded.	Fine subangular blocky	<10
				Medium subangular blocky	10-20
				Coarse subangular blocky	20-50
				Very coarse subangular blocky	>50
		C.	Granular-spheroidal- characterized by rounded vertices.	Fine granular	<2
				Medium granular	2-5
				Coarse granular	5-10
3	Platelike-soil particles arranged around a horizontal plane and generally bounded by relatively flat horizontal surfaces.	A.	Platy-horizontal planes more or less developed.	Fine platy	<2
				Medium platy	2-5
				Coarse platy	>5
4	Prismlike-soil particles arranged around a vertical axis and bounded by relatively flat vertical surfaces.	A.	Prismatic-vertical faces. Fine well defined and edges sharp.	Fine prismatic	<20
				Medium prismatic	20-50
				Coarse prismatic	50-100
				Very coarse prismatic	>100
		B.	Columnar-vertical edges near top of columns not sharp. (Columns may be flat-topped round-topped, or irregular).	Fine columnar	<20
				Medium columnar	20-50
				Coarse columnar	50-100
				Very coarse columnar	>100

Table 5.5. Soil structure classes (quoted from Research Branch, 1976: http://sis.agr.gc.ca/cansis/glossary/table-3.html).

clay loam, clay loam, sandy loam, sandy clay loam and loamy sand.

- Stoniness

Stoniness is the relative proportion of stones in the soil. The quantity of stones can restrict plant life in some cases or facilitate water infiltration in others. Stoniness was measured as a percentage in each soil horizon. In Calakmul, the range varies between 3 to 50% (excluding the C horizon).

- Structure

The accommodation of the soil aggregates, called peds, form what is called soil structure. The ways in which water, air, heat, and nutrients pass through the soil is very important for the proliferation of both plants and microorganisms (Eyhorn et al., 2002). The structure of Calakmul soils is basically blocky subangular, although there are some single grained and amorphous soils. The classification of soil structure is shown in Table 5.5.

- Humidity

Water retention is indispensable for agriculture and plant growth in general. The ways in which soils preserve humidity is of vital importance for plant life. A balance in humidity is crucial, since extremes are not conducive to fertility. In this study a qualitative humidity index was used: dry, humid and aqueous. Dry soils had good drainage and kept the same color after laboratory analyses. In contrast, aqueous soils had poor drainage and water could be observed in the profile. Humid soils absorbed water and their weight was decreased when the samples were dried out. The majority of Calakmul soils were humid.

Chemical Properties

The chemical properties of soil depend on the mineral composition of the matrix and the way it reacts to the physical environment. Comprehension of these properties leads to a more complex and complete understanding of soil fertility and formation. The chemical properties of the soil are described below. The complete results of these analyses can be appreciated in Appendix 3.

- Organic Matter

Soil can be viewed as a microcosm with thousands of organisms living in it. The presence of both macroscopic (for example, earthworms, spiders, snails, ants, and ticks) and microscopic (such as bacteria, algae, protozoon, or fungus) entities can be observed. These organisms decompose organic matter, create humus, and combine organic matter with soil particles forming soil structure. Some organisms make tunnels helpful for plant roots and conducive to soil aeration, assist in the release of nutrients, and even control some pests. Furthermore, their excrements contain nutrients that are used by plants (Eyhorn et al., 2002).

The quantification of organic matter is useful for inferring the life activity of various organisms within each horizon and its subsequent suitability for supporting plants. The organic matter percentage in Calakmul soils ranges from 6 to 32%. This elevated percentage is not surprising in a biosphere reserve.

- pH

The pH is a measure of the hydrogen ions quantity present in the soil. The pH scale ranges from strongly acidic (pH of 0) to strongly alkaline (pH of 14). The majority of plants and important microorganisms grow in the 6 to 7.5 pH range, although some plants are adapted to their natural conditions that can be more alkaline or acidic. The pH ranges are indicated in Table 5.6. Calakmul soils are moderately alkaline and neutral, excluding C horizons that are strongly alkaline.

Category	Range
Strongly acid	below 5.1
Moderately acid	5.2 – 6
Slightly acid	6.1 – 6.5
Neutral	6.6 – 7.3
Moderately alkaline	7.4 – 8.4
Strongly alkaline	above 8.5

Table 5.6. The pH categories for soils (Marx et al., 1999).

- Calcium Carbonates

Although calcium carbonates are intrinsic components of limestone rocks, their relative amounts in each horizon affect plants in particular ways. A good reference of the extent of carbonates can be obtained by adding an acid to

Description	%$CaCO_3$	Audible Effects	Visible Effects
Non-calcareous (<0.5%)	0.1	none	none
Very slightly calcareous (1 to 2%)	0.5	little audible	none
Slightly calcareous (2 to 5%)	1.0	moderately audible	detectable effervescence
Moderately calcareous (5 to 1%)	2.0	clearly audible	slightly effervescence
Calcareous (0.5 to 10%)	5.0	easily audible	moderate effervescence
Highly calcareous (10%)	10.0	easily audible	strong effervescence

Table 5.7. Qualitative analysis of carbonates (Based in León, 2003).

each sample. A qualitative result is generated following the code observed in Table 5.7. As expected, Calakmul soils are calcareous and highly calcareous.

- Soluble Salts

Salinity is the presence of salts in the soil. When soluble salts are high, plant growth is affected. Salinity in the soil is measured by soil conductivity. The following categories of soluble salts are seen in Table 5.8. The majority of Calakmul soils present low conductivity, with a range of 640 to 1280 ppm salts.

Conductivity	mmhos/cm*	ppm salt**
Low	<1.0	<640
Medium	1 – 2	640-1280
High	>2	>1280

*Mmhos/cm ~ decisiemen/m
**Multiply mmhos/cm by 640 to obtain ppm salt

Table 5.8. Conductivity levels for soils (Based in Marx et al., 1999).

- Phosphorous

Phosphorous is an essential element for plants. It can be found in living organisms, soil and water. It is a vital component for many physiological processes related to proper energy utilization in both plants and animals (Sharpley et al., 1994). The levels of phosphorus in soils are shown in Table 5.9. The phosphorous levels in the soils of Calakmul are very low to medium.

Phosphorous	Mehlich-3 Soil (Lbs/acre)	Test Value ppm
Below Optimum (very low)	0–24	0–12
Below Optimum (low)	25–45	13–22
Below Optimum (medium)	46–71	23–35
Optimum (high)	72–137	36-68
Above Optimum (very high)	>138	>69

Table 5.9. Phosphorous levels categories for agriculture (based in Heckman, n.d.).

- Cation Exchange

The quantity of cations that a soil holds, specifically, K, P, Mg, Ca, H and Al, is called the cation exchange capacity (CEC). This number is an index of the capacity of the soil to retain nutrients and as a consequence, serves as a measure of actual soil fertility. In order to grow economic crops, the concept of an optimal cation ratio has been proposed for fertilizer recommendations, essential for large-scale agriculture. The ratios are the "ideal" proportion of each element in a specific land for a particular crop. But even some soil scientists state that there is very little hard research or demonstrable evidence to demonstrate that these ratios have either a positive or negative effect on crop production (Rehm, 1994). In any case, these ratios change according the proposed crops, terrain, scale and of course are time-sensitive. Since in this project there is not a valid parameter for comparison, these costly analyses were left for future research, after generating enough data for a solid comparison. However, the phosphorus content analysis was done as an indirect measure of fertility.

- Fertility

Fertility is a measure of the capability and/or suitability of the soil for providing the necessary conditions conducive to vegetative life cycles. All the cited properties, as well as human influence, contribute to soil fertility. In the earth sciences fertility is measured by considering the availability of cations and their interchange capability in order to be assimilated by plants. In this study and only as a rough index, the phosphorus level was measured.

The Soils of Calakmul

The soils of Calakmul are formed by soft limestone (with a 90% level of calcium carbonate) with small fractions of iron, silica and aluminum (Morales, 1999). After following the mentioned procedures, four types of soils and their distribution were identified. They have been identified in previous and current FAO classifications, as well as in Maya taxonomy (See Table 5.1).

- Lithosols (I), Lithic Leptosol (Lpq) or *Tsek'el*

Lithosols are non-evolved soils with less than 10 cm of depth. The FAO soil map (FAO, 1974) describes them as very shallow soils, over hard continuous rock or weathered limestone (with $CaCO_3$ levels above 40%). They were observed only in the top or steep sides of the hills, and next to rock outcrops (Figure 5.4). The name Leptosol comes from the Greek *leptos* or thin. In Maya, *tsek'el* means very stony earth (Bastarrechea et al., 1992). In Calakmul, Lithic Leptosols with less than 8 cm of depth were observed. They exhibited an A black or very dark grayish brown color horizon, weak structure, abundant stones, high organic matter content, and sandy loam texture. Rocky outcrops were observed next to these soils. The vegetation associated with these soils is medium semi-evergreen forest.

Figure 5.4. Lithosol. Pit D. Notice the rocky outcrop and the herbal cover.

- Rendzinas (E), Rendzic Leptosol (Lpk) or *Pus lu'um*

Rendzinas are alkaline soils that present a dark mollic A horizon, with more than 10 cm of depth. They are developed above highly calcareous rock or soft parent material. The FAO map (FAO, 1974) describes them as shallow soils with dark humic topsoil. In Calakmul they appear in hills and micro-valleys (Figure 5.5). They have a dark gray and gray A horizon with high organic matter content. This horizon has an elevated biological activity due to multiple organisms that live in the soil and rapidly decompose it. This soil is well aerated, has abundant to regular stoniness, shallow depth, and a silty clay loam texture. In Soil Taxonomy classification these soils correspond with Rendolls.

Figure 5.5. Rendzina. Pit N. The white rock is the beginning of the parent material.

Rendzinas detected in the Biosphere are divided according to Maya soil classification in three groups: black soils with stones (*box lu'um*), gray soils with weak consistence and few rocks (*pus lu'um*), and brown soils with granular structure (*chi'ich lu'um*) (Morales and Magaña, 2001). In Calakmul, we observed *pus lu'um* in micro-valleys and *box lu'um* on hillsides. Vegetation associated with Rendzinas is characterized as medium semi-evergreen forest.

- Calcaric Vertisols (Vc), Calcic Vertisol (Vk) or *Ya'ax Hom*

These soils are dark soils with an over 30% content of montomorillonith-rich clay. As a result of exposure to water and subsequent drying, the clay minerals expand and contract. This phenomenon is observed through cracks at a depth of between one to fifty centimeters when soils are dry, gilgai microrelief, intersecting slickensides, or wedge-shaped structural aggregates tilted at an angle from the horizon (information from FAO web site http://www.fao.org/ag/agl/agll/prosoil/verti.htm, accessed 5th May 2005). These Vertisols showed a calcic horizon. A calcic horizon is formed with more than 15% of calcium carbonates aggregates above non-calcic strata (León, 2003). In Soil Taxonomy classification these soils correspond with Vertisols.

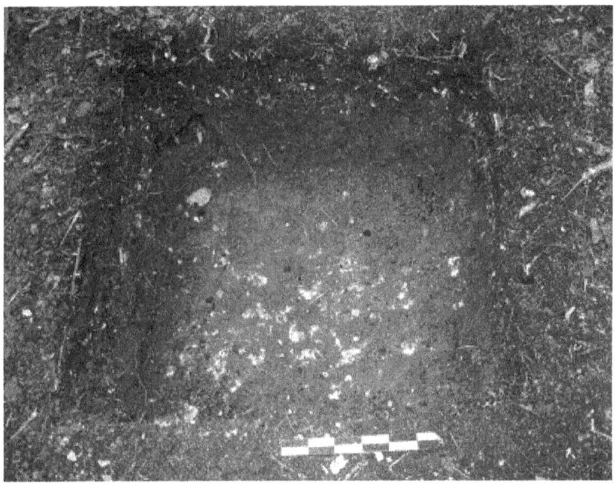

Figure 5 .6. Calcic Vertisol. Pit J. Notice the superficial calcic horizon.

In humid environments, Vertisols show a viable potential for agricultural production. However, their clayed matrix makes them difficult to work because of their stickiness, weight, and rigid solidity. The principal problems related to their management are flooding, elevated contents of Mg and Ca, high pH values, and low levels of N and phosphorus. Their advantages are that they are chemically rich, capable of sustaining continuous cropping, and that they do not necessarily require a rest period because pedoturbation brings subsoil to the surface.

However their productivity is not elevated. In addition, the structural stability of Vertisols is low and it is easily altered by water erosion. On gentle slopes, contour cultivation with a ground cover crop is possible. When terracing, a drainage method should be planned in order to canalize water (FAO web site: http://www.fao.org/ag/agl/agll/prosoil/verti.htm, accessed 5th May 2005 and data from the site of edaphology: http://www.soils.org/sssagloss/cgi-bin/gloss_search.cgi?QUERY=Vertisol&SOURCE=1, accessed 5 May 2005). The Calcic Vertisols of Calakmul (Figure 5.6) are black or grayish brown, exhibit moderately deep soil depth and poor drainage, are rife with rocks, show surface cracking, have high organic matter content and a clay loam texture.

- Vertic Gleysols (GLv) or *Ak'alche*

Gleysols are soils with gray colors, often with mottles due to wetness within 50 cm of depth (FAO, 1974). The gleyic soils have poorly-drained dark-gray A horizons, with moderately high organic matter, underlain by mottled gray or brownish gleyic mineral horizons. In this kind of hydromorphic soil organic decomposition is anaerobic, since they are found in flooded landscapes. These soils are not good for life. In Soil Taxonomy classification, these soils correspond to Aquents and Aquepts.

Figure 5.7. Vertic Gleysol. Pit K. Notice the hard clayed matrix.

Vertic Gleysols in Calakmul are well-developed soils located in flat areas that are flooded seasonally, although the water does eventually drain into non-permanent currents, into the *corriental* or evaporates (Figure 5.7). Mottles were not observed in the Vertic Gleysols of this area. A vertic horizon is inferred because of elevated clay content and the cracks, which appear during the dry season. These soils have poor drainage, shallow soil depth, high organic matter content, almost no stones, and a clayish texture. Their typical vegetation is low semi-evergreen forest.

The Soil Map of Calakmul

After soil identification, a map of the soils of Calakmul was generated following the FAO soil classification (1974) and INEGI edaphological units (INEGI, 1998). The resulting soil information was digitized with Autocad 2004 and exported to the software programs ArcGIS 8.3 and Autodesk Map 6 to generate the soil map itself (Figure 5.8).

The soils of Calakmul were identified as single soils and associations. In Calakmul different soils form a complex mosaic, and the results are reported correspondingly. The identification of these associations coincides with those reported by Morales (1999). In Calakmul, over an area of 64.60 km^2, three kinds of soils or soil associations were observed: Lithosols, Rendzinas-Calcic Vertisols and Calcic Vertisols-Vertic Gleysols. Their quantities and percentages are expressed in Table 5.10.

Soil	km^2	% of 64.60 km^2
Lithosols	0.44	0.70
Rendzinas-Calcic Vertisols	23.24	36.0
Calcic Vertisols-Vertic Gleysols	40.92	63.3

Table 5.10. Calakmul soil associations' quantity and percentage.

Lithosols are located in steep areas. They surround the tri-lobed promontory and some sections that divide the hills and elevations from the *bajos*. They are found in the ravines and also on some rock outcrops that, due to reasons of scale, are not evident in this map. Although they are not deep, in some places thin vegetation or young plants cover them. They are not as abundant as one may think; they cover less than 1% of the study area.

Rendzinas-Calcic Vertisols are found in the elevated areas of the Calakmul promontory and associated hills. Particularly, Rendzinas are observed in the hills and micro-valleys, usually next to areas that evidence human inhabitation. This soil association is seasonally inundated and it is easy to identify it because of the big cracks that develop during the dry season. Although their vegetation is classified as medium semi-evergreen forest, it is readily apparent that more and taller trees grow in areas with a predominance of Rendzinas.

Finally an association between Calcic Vertisols and Vertic Gleysols was observed. These soils are located in the *bajo* or flat areas. Calcic Vertisols were seen on the edges of the *bajos* and *aguadas*. Vertic Gleysols, too, were evident in the *bajos* and on the plains, but interestingly, also on surfaces that demonstrate considerable water movement and long-term inundation. Consequently, these soils appeared in the *corriental*, in non-permanent currents and in channels. It is important to note that at times it was difficult to differentiate one soil from another because of the presence of transition soils (which share some features of both Vertisols and Gleysols).

Although it seemed that many kinds of soils would be present in Calakmul, only four of them were identified. What appears to be of greater complexity is their distribution, since they present themselves graphically as a mosaic. A much more detailed field reconnaissance would be necessary to identify the precise location of each soil. However, their resultant positioning as expressed in maps are sufficiently suitable in terms of discussing agricultural land potential and soil use.

However, the soil or the vegetation maps per se are not detailed enough to provide information that suggests land use, or to support any extant interpretation thereof. This information should be "translated" in terms of a suitability classification to define the relevant land use alternatives (Breimer et al., 1986). The next chapter will discuss how the Maya could have managed their landscape according to the examination of the land potential.

Land Evaluation

Land evaluation is the process of estimating the potential of land for alternative uses, such as arable farming, grazing or forestry (Dent and Young, 1981). This kind of assessment is currently performed in many countries of the world in order to take full advantage of the most productive developmental options. Although land evaluation represents actual interests, it frequently involves change and its effects on the use of land, and in some cases, changes in the land itself (FAO, 1976). For these reasons, land evaluation is valuable for archaeologists. Since any landscape is a snapshot which shows the amalgamation of present and past conditions, it is possible to infer the historical influence of humans through the observed changes over time.

As part of a land evaluation study with an archaeological focus, some questions can be posed (modified from FAO, 1976):

1. Which were and are the most economically and socially relevant potential uses?
2. What are the benefits of each form of land use?
3. How was the land managed?
4. What improvements were and are possible in management practices?
5. Which of these uses offers the greatest possibilities of sustained production?
6. What recurrent inputs are necessary to bring about the desired production and minimize adverse effects?

As a result of the land evaluation process, two or more potential uses for each specific area of study should be suggested, and beneficial and/or adverse consequences also considered. This provides a more flexible framework in which to discuss the alternatives and ultimately select the most appropriate option, bearing in mind the ancient context of the study area.

The specific objective in evaluating a terrain is to guide decisions on human land management by suggesting the most beneficial use, and at the same time, to ensure sustainability by preserving resources for the future (FAO, 1976). In archaeology, it is not necessary to plan for a specific contemporary land use (as pasture or grain crops) in the study site, rather to observe the characteristics of the area and the potential land use that will not deplete environmental resources, at least for long periods. For example, we know that the Maya practiced agriculture for at least 3000 B.C. in the Maya lowlands (Pohl, 1990). It is therefore quite obvious that their subsistence methods were sufficient to satisfy their primary necessities for many centuries. This fact suggests that their procurement methods were sustainable in the long-term.

One of the byproducts of the process of land evaluation is the land capability classification. Land capability classification, originally proposed by Klingebiel and Montgomery for North American farmers, is an interpretative grouping of functionally similar soils that require comparable management measures (Klingebiel and Montgomery, 1961). Capability refers to the range and intensity of possible uses in a specific locale. It implies versatility in terms of agricultural potential but not necessarily the best areas for production (FAO, 1989). Likewise, land suitability is the suitability and fitness of land for any particular kind of use (FAO, 1993). Hence, this work treats both terms as synonymous.

The North American approach to conducting land evaluation (Klingebiel and Montgomery, 1961) consists of isolating the relevant land attributes and dividing the range of values or intensities associated with each attribute into eight different classes. Only classes I to IV are useful for agricultural purposes. This method has previously been used for archaeological research in the Maya area (e.g. Fedick, 1995; Geovannini, 1998). However, this land classification is biased towards contemporary agricultural practices. In effect, the method has the tendency to search for the optimal places for *mechanized* agriculture, probably due to the fact that it was developed for North-American landscapes. For this reason, a method for land evaluation more related to the Maya area was applied.

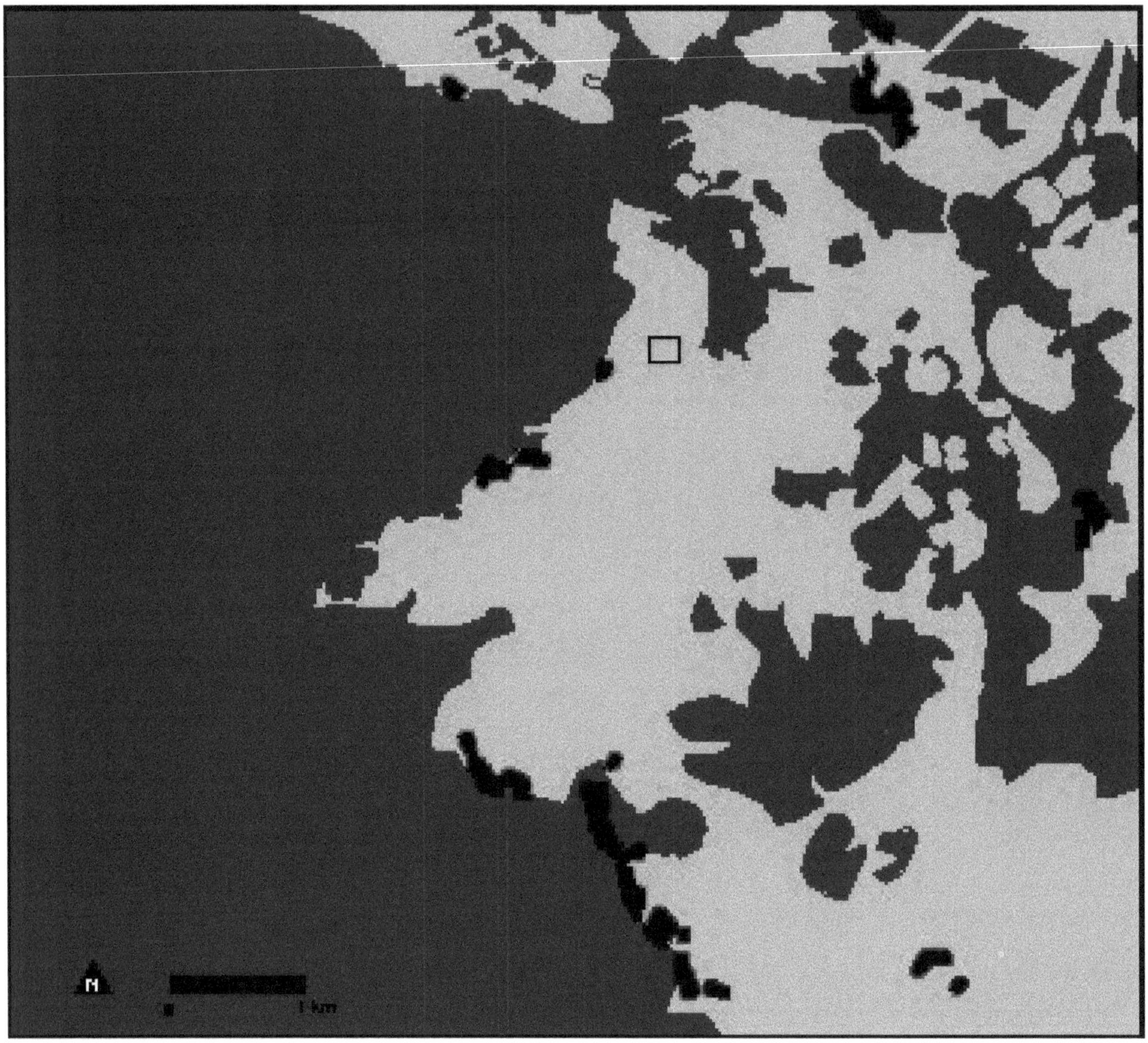

Figure 5.8. Map of the soil associations of Calakmul. The square represents Calakmul civic-ceremonial center. Lithosols are shown in black; Rendzinas-Calcic Vertisols in light gray and Calcic Vertisols-Vertic Gleysols in dark gray.

Procedures for Land Evaluation in Calakmul

In this study, a specific land capability classification criterion for the current land potential in the humid tropics was used. This system has been used successfully in several tropical countries in the developing world (FAO, 1989; 1999). First, the landscape is observed, described and several pits are made according to the variation of the setting. Second, each pit is described according to relevant classification criteria (with the aim of identifying the soils). These criteria are: slope or inclination; soil depth or effective depth (the length plant roots can penetrate); stoniness, as estimated in the upper 60 cm; fertility, measured through available phosphorous; texture, that is, the particle size of the soil (USDA, 1962); erosion and drainage, according to soil humidity and to the depth below the surface at which more than 10 percent of gray mottles are first encountered. Each pit is evaluated with these criteria to assess its correspondent class. For the criteria description and ranking, FAO procedures are followed (FAO, 1989; 1999). It is important to mention that in this method fertility is assessed by the sum of exchangeable bases (calcium, magnesium, potassium and sodium), but only phosphorous levels were used as the pertinent index. Finally, with the resulting evaluation classes for each plausible land utilization type, a potential land evaluation map for Calakmul was proposed.

It goes without saying that a specific terrain does not possess an intrinsic value in terms of land evaluation. The sum of all criteria suggests only a potential response to a query regarding the land. Each investigation has different goals and methods. The basic question in this case is related to the capability of the land for sustaining the ancient Maya population. It is obvious that a considerable number of people lived in Calakmul from the Preclassic period through the Late Classic times, so how they managed their environment in order to subsist, survive, and thrive is a critical question.

The decision-making process for identifying optimal land use category is shown in Table 5.11. The first step was to consider the slope percentage, and then the effective depth, and finally stoniness for each kind of soil observed in Calakmul. Each soil is indicated with a particular color in each factor column. Since Rendzinas were found both in steep and in flat areas, two different evaluation criteria were used.

After obtaining these results with the methodology of FAO (1989), the fertility, drainage and texture criteria were added in order to emphasize the plausible limitations of the terrain and to refine the classification. The results are shown in Table 5.12.

Having observed Tables 5.11 and 5.12, the optimal land category options can now be discussed. For example, Lithosols and Rendzinas located in the valley can be used as natural forests. Considering the archaeological context of Calakmul, this terrain coincides with the primary built area of the site. The main restrictions of these soils are their shallow depth and low fertility.

Rendzinas located in the hills are more fertile and hence their optimal use is as pastureland. Since livestock was not raised in Mesoamerica, this terrain could also have been used as forest. Calcic vertisols exhibit the greatest number of potential utilizations. They can be managed to cultivate grain crops, timber trees, fruit forage and fuel-wood trees, tuber crops horticultural crops, or simply as pasture. They have the best fertility indexes in the area, and their restrictive properties apply mainly during the dry season. In many parts of the world these soils are cultivated extensively.

Vertisols are found in the semi-arid tropics, mainly in Africa (Central Sudan, South Africa, Ethiopia, and Tanzania), Asia (the Deccan plateau of India), Australia and America. In areas where irrigation a viable option, or in more humid environments, the use of vertisols is more feasible, and their potential for sustained agricultural production is significantly increased (http://www.fao.org/ag/agl/agll/prosoil/verti.htm, accessed September 8th, 2004). Finally, some areas with Vertic gleysols can be used for cultivating grains (and as pasture). Their main limitations are susceptibility to inundation and high clay content.

However, identifying the primary potential land uses is only the first procedure in land evaluation. The next stage is to propose types of land utilization in which these broad, general uses are refined and described in greater detail. A hypothetical example would be rainfed agriculture based on maize subsistence with intensive chili cultivation.

The suitability classification recognizes orders, classes, subclasses and units. The orders are simple: suitable (S) or non-suitable (NS) for land use. Then, the classes are selected according to the characteristics shown in Table 5.13.

The next step is to identify subclasses that reflect land limitations. Letters implying potential hazards are put into classes, for example S3e (e signifying erosion). The S1 Class, however, does not have subclasses. All this information is transferred to suitability maps where the current or potential use is assessed (FAO, 1976). The factors for weighing a suitability map can be quantitative or qualitative.

Slope %	Effective depth (cm)	Stoniness (%)	Optimal land use category
60-70	<25		F
	25-50	>25	F
		<25	TT
	50-100	>50	F
		<50	TT
50-60	<25		F
	25-50	>50	F
		25-50	F
		<25	TT
	50-100	>75	F
		50-75	TT
		25-50	TT
		<25	FT
20-50	<10		F
	10-25	>25	F
		<25	P
	25-50	>50	F
		25-50	P
		<25	G,P
	50-100	>75	F
		50-75	G,P
		25-50	G,P,FT,TT
		<25	G,P,FT,TT
10-20	<10		F
	10-25	>25	F
		<25	P
	25-50	>50	F
		25-50	P
		<25	G,P
	50-100	>75	F
		50-75	G,T,P
		<50	G,T,P,FT,TT
	>100	>75	G,T,P
		50-75	G,T,P
		<50	G,T,P,FT,TT
0-10	<10		F
	10-25	>25	F
		<25	P
	25-50	>50	F
		25-50	P
		<25	G,P
	50-100	>75	F
		50-75	G,H,T,P
		<50	G,H,T,P, FT, TT
	>100	>75	G,H,T,P
		50-75	G,H,T,P
		<50	G,H,T,P, FT, TT

Table 5.11. Rating factors for the identification of optimal land use categories. F is forest; G, grain crops; P, pasture; TT, timber trees; FT, fruit forage and fuel-wood trees; T, tuber crops; H, horticultural crops. Lithosols have 50-60 slope %; valley Rendzinas an effective depth between 10 and 25 cm; hill Rendzinas, between 25 and 50 stoniness %; Vertic gleysols less than 25 stoniness % and Calcic vertisols less than 50 stoniness % (FAO, 1989).

Soil	Land unit	Slope %	Effective depth cm	Stoni-ness %	Ferti-lity	Drain-age	Tex-ture	Optimal land use category
Lithosol	steep hills	57.7	6-10	30	very low	good	sandy loam	F
Rendzina	hills	3.4-17.6	25-30	25-30	very low	good	silt	P
Rendzina	valley	0	17-29	25-40	low to very low	good	silt sandy	F
Calcic Vertisol	valley or plain	0-5	33-83	3-25	mediu m to very low	good	clay loam	G,H,T,P, FT, TT
Vertic Gleysol	plain	0	26-27	3-15	low to very low	poor	clay	G, P

Table 5.12. Rating factors for the identification of optimal land use categories. F signifies natural forest; G, grain crops; P, pasture, natural or improved; TT, timber trees; FT, fruit forage and fuel-wood trees; T, tuber crops; H, horticultural crops (based on FAO, 1989 and 1999).

Class S1 Highly Suitable	Land having no significant limitations to sustained application of a given use, or only minor limitations that will not significantly reduce productivity or benefits and will not raise inputs above an acceptable level.
Class S2 Moderately Suitable	Land having limitations, which in aggregate are moderately severe for sustained application of a given use. The limitations will reduce productivity or benefits and increase required inputs to the extent that the overall advantage to be gained from the use, although still attractive, will be appreciably inferior to that expected on Class S1 land.
Class S3 Marginally Suitable	Land having limitations, which in aggregate are severe for sustained application of a given use and will so reduce productivity or benefits, or increase required inputs, that this expenditure will be only marginally justified.
Class Sc Conditional Suitability	Small areas of land, within the survey area, that may be unsuitable or poorly suitable for a particular use under the management specified for that use, but suitable given that certain conditions are fulfilled.
Class N Not Suitable	Land that cannot support the land use on a sustained basis, or land on which benefits do not justify necessary inputs.
Class N1 Currently Not Suitable	Land having limitations which may be surmountable in time but which cannot be corrected with existing knowledge at currently acceptable cost; the limitations are so severe as to preclude successful sustained use of the land in the given manner.
Class N2 Permanently Not Suitable	Land having limitations which appear so severe as to preclude any possibilities of successful sustained use of the land in the given manner.

Table 5.13. Suitability classes of land evaluation (quoted from FAO, 1976: http://www.fao.org/documents/show_cdr.asp?url_file=/docrep/X5310E/X5310E00.htm).

Suitability Classes in Calakmul

The first and most obvious land utilization type for the Calakmul area is as a forest, taking advantage of the land through the exploitation of valuable wild species for food and medicine and collecting timber for fuel or construction. Virtually, the region can be exploited as a forest, being the inundation of large areas during the rainy period the principal constraint for access.

But considering the ancient context of Calakmul, with thousands of individuals during the Classic period, an agricultural system was practiced for sure. Turner and colleagues (2003) have proposed two main agricultural systems for use in the Maya area: Swidden *milpa*, with maize based systems in which fallow length exceeds that of cultivation (1:4) and intensive swidden maize based systems in which the length of fallow and cultivation are equal (1:1). However, direct evidence of the slash and burn technique has not been obtained, although it has been well documented in many contemporary Mesoamerican areas.

In Calakmul, two land utilization types are analyzed as part of this study: Rainfed *milpa* agriculture with two crops per year and rainfed *milpa* agriculture with one crop per year. *Milpa* means a fallow-based cultivation system employing slash-and-burn techniques that is practiced at present in Maya area. This system is described with more detail in the next chapter.

- Rainfed Milpa Agriculture with Two Crops per Year

An inherent pre-condition for this type of land utilization is almost continuous rainfall that would allow sustained land use year-round, without providing extra water. In and around Calakmul, there is indeed year-round precipitation, as shown in Figure 3.4. The dry period begins in November and lasts until April. In May, the rainy season begins and increases in intensity until June, when rains decrease slightly for one or two months. After mid to late July rainfall continues to increase in intensity until it peaks in October. After October the rains decrease significantly for a period of six months.

Considering the previous suitability classifications and this proposed land utilization type, a map was generated using Autodesk Map 6. The vegetation, soil, and slope vector layers were overlaid through a Boolean operation and a landscape with different six different characteristics was obtained. This terrain was evaluated according to its suitability for two crops each year based on available data regarding average rainfall and terrain humidity in the dry season. This map is shown in Figure 5.9.

Figure 5.9 shows that a small area (6.6 km^2) is moderately suitable for yielding two crops per year, with some limitations associated with inundation and clay content. This area (S2ic, with gray color) is unique because it represents the transition between hills and *akalche*. It receives rainwater but does not retain it for long periods; rainwater gradually recedes into lower, flatter areas. Calcic Vertisols are the dominant soil type in this area. The main restriction for this surface is inundation during heavy rains, as it could conceivably be inundated if more level, lower-lying terrain reaches its saturation point. In addition, the soils of this area also have high clay content that makes them difficult to work.

The white areas in the map (58 km^2) are those that are marginally suitable land (S3sic), prone to inundation, with shallow soils and clay constraints. Such terrain is difficult to cultivate, although not impossible, for three reasons. The first is that its shallow soils, while allowing growth, restrict the fullest development of the plants. The second is that some of these areas are subjected to continuous flooding that impedes agriculture for some months. Lastly, some terrain patches have soils with a high clay content that also restricts plant development.

Finally, the narrow extension black areas in the map (0.044 km^2) are unsuitable (N1) for this type of land utilization, since cultivating in shallow soils with very steep slopes presents far too many constraints for agriculture.

A calendar for this type of rainfed *milpa* subsistence agriculture is presented in Table 5.14. The first crop may follow the normal 8-9-month cycle; the second, only a 5-6-month cycle. Agriculturalists would likely have chosen crop varieties suitable for first and second year cultivation as well as short and long cycle. This topic will be discussed in the next chapter.

- Rainfed Milpa Agriculture with One Crop per Year

This is a kind of land utilization terrain for only one agricultural cycle per year, because there is insufficient rainwater for another (its scarcity would affect the plants physiological processes). In dry years (with insufficient rain) some kind of irrigation would have been necessary in order to obtain satisfactory yields. The calendar for such a management scheme would be the first year part as shown in Table 5.14.

Again, after considering the previous suitability classifications and this proposed land utilization type, a map was generated using Autodesk Map 6. The vegetation, soil, and slope vector layers were overlaid through a union Boolean operation. Also this terrain was evaluated on available data regarding average rainfall and terrain humidity during the dry season.

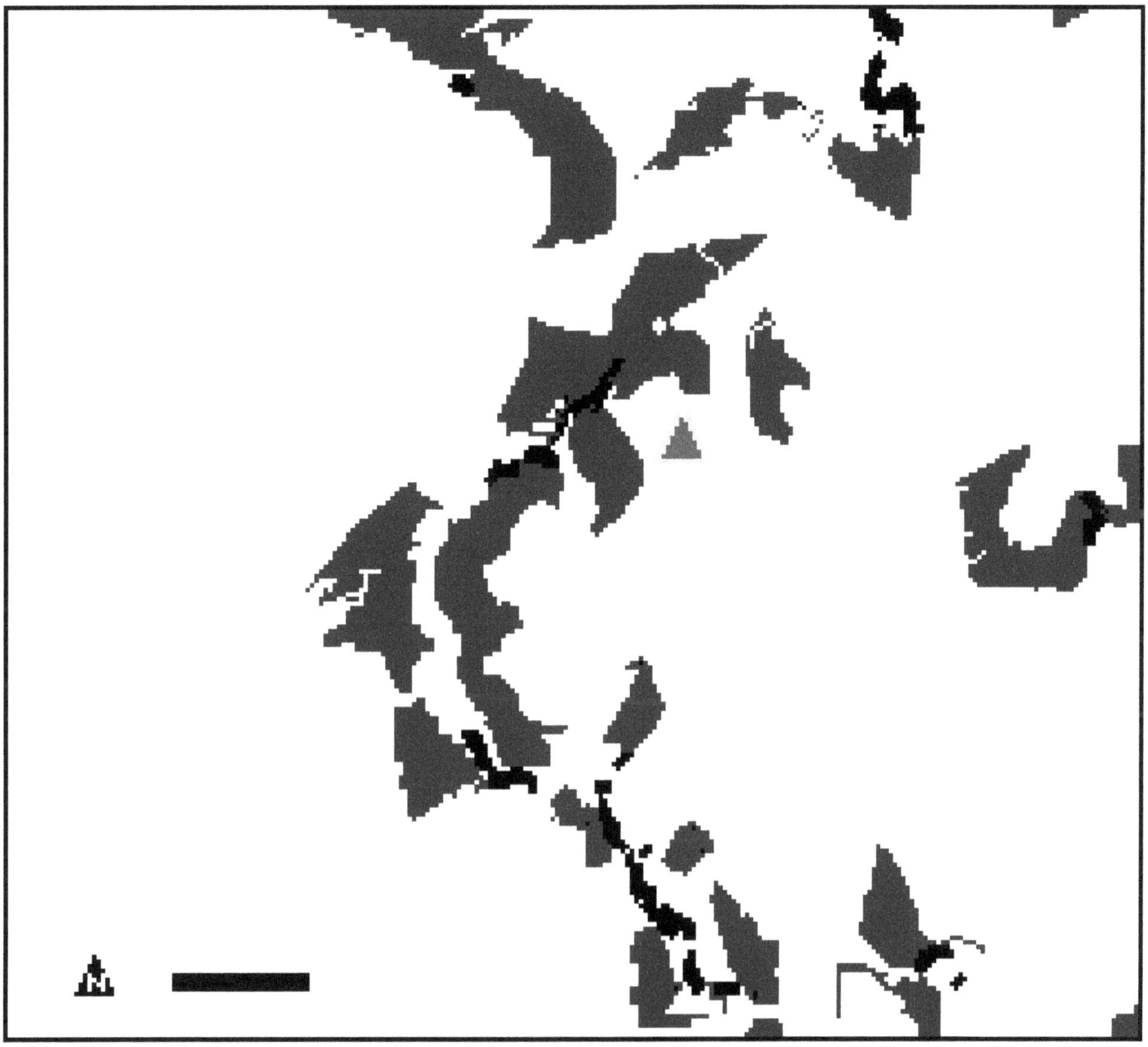

Figure 5.9. Map of suitability classes for rainfed milpa agriculture with two crops per year. Black area represents N1 (non-suitable) terrain; white area, S3sic (marginally suitable with shallow soils, inundation and clay constraints) and gray area S2ic (moderately suitable with inundation and clay constraints). The scale measures 1 km and the civic-ceremonial center is marked by the gray triangle.

FIRST CROP											
Jan	**Feb**	**Mar**	**Apr**	**May**	**Jun**	**Jul**	**Aug**	**Sep**	**Oct**	**Nov**	**Dec**
		burning		**sowing**				**bending**			
					hand weeding						
harvesting								**harvesting**			
SECOND CROP											
	bending							**sowing**			
harvesting									**hand weeding**		

Table 5.14. Calendar for cultivating milpa, two crops each year (based in Arias, 1994 and two traditional Yucatan farmers personal communication, 2005).

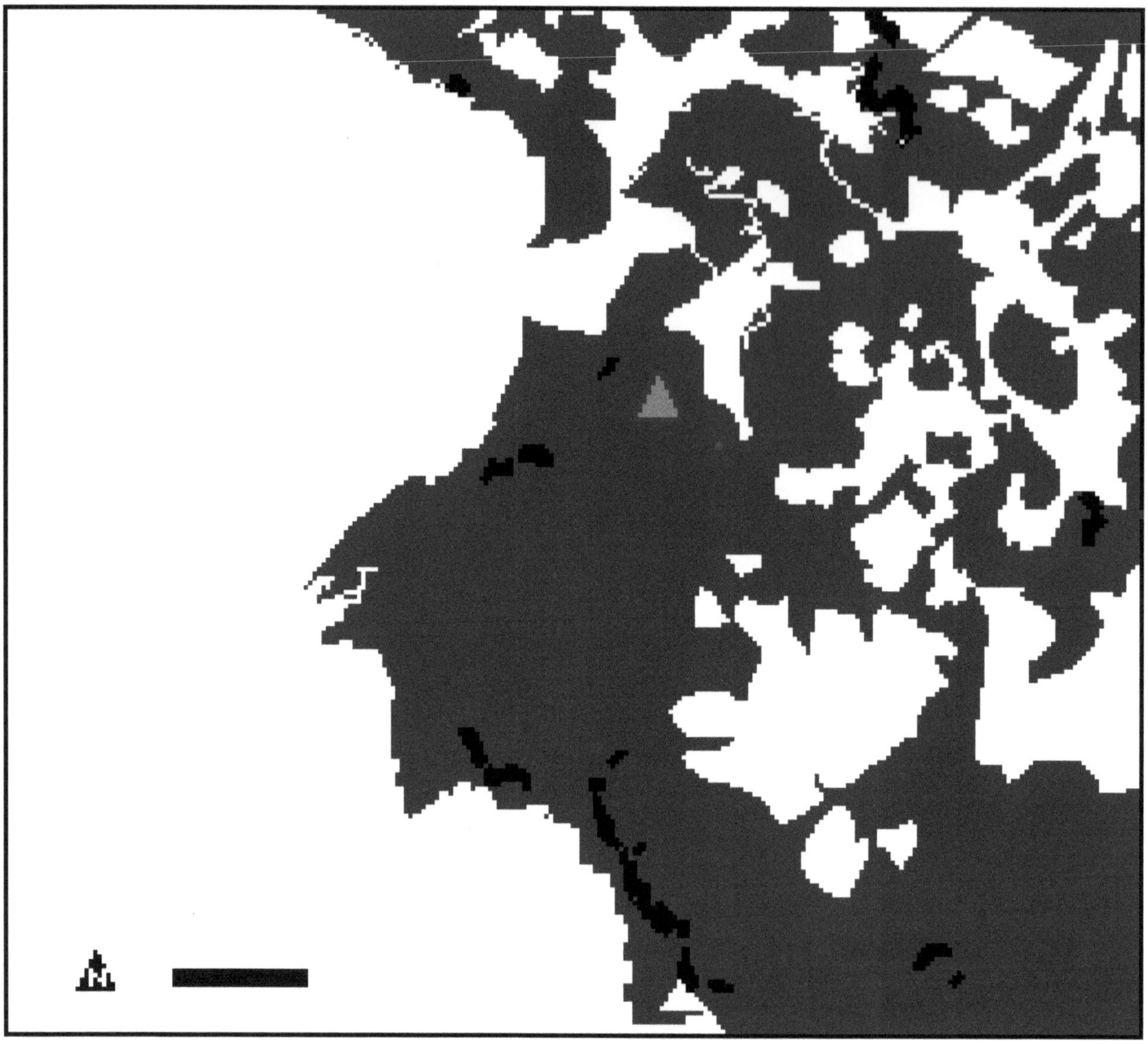

Figure 5.10. Map of suitability classes for rainfed milpa subsistence agriculture with one crop per year. Black area represents N1 (non-suitable) terrain; white area, S3ic (marginally suitable) and gray area S2si (moderately suitable). The scale measures 1 km and the civic-ceremonial center is marked by the gray triangle.

If terrain is cultivated with one annual maize crop, the suitable land would correspond to that illustrated in Figure 5.10. As this map shows, a much larger area (of 28.01 km^2, colored in gray) could have thrived with one-crop rainfed agriculture. This moderately suitable land utilization type with shallow soils and inundation constraints (S2si) represents about the half of the area. The soils in the area are Rendzinas and Calcic Vertisols, some of them very shallow, and others susceptible to saturation and waterlogging.

The portions shown in white represent marginally suitable areas with inundation and clay restrictions (S3ic). This area (36.54 km^2) could also be cultivated if flooding was prevented through the modification of the terrain (such as raised fields connected by channels). And finally, the black areas represent terrain that is unsuitable for this type of land utilization (0.044 km^2). Again, cultivating in shallow soils with very steep slopes is hardly adequate for this kind of agriculture.

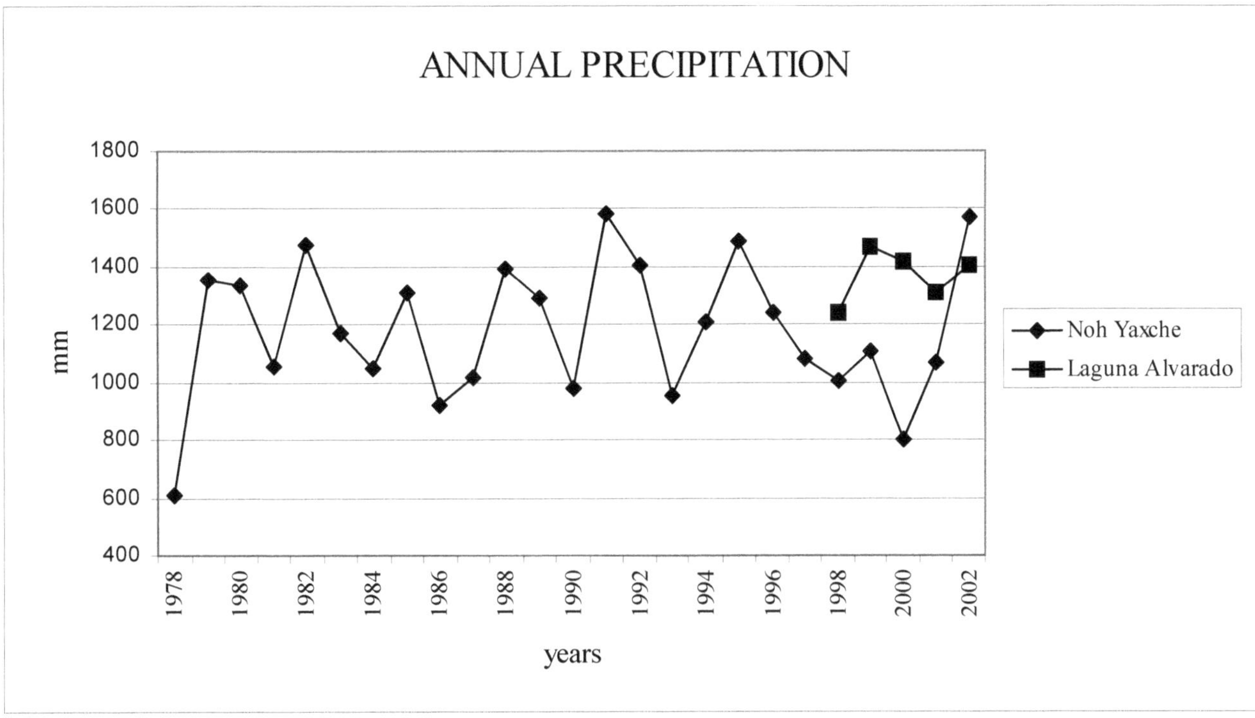

Figure 5.11. Average of total annual precipitation in Noh Yaxche (24 years) and Laguna Alvarado (5 years) stations, located in Campeche, near the Calakmul Biosphere Reserve (Comisión Nacional del Agua, 2005).

Calakmul Swidden System Yield

In order to consider the suitability of Calakmul land for *milpa* swidden system, the daily average maize consumption and the average maize yield per area should be considered. After revising several studies the best number for maize daily consumption for people of all ages is 0.77 kg (Cowgill, 1962:277). Hence, the annual average annual consumption of maize is 281 kg per person. In the same way, an accepted number for annual average maize yield is 1200 kg/hectare (Murtha, 2002).

If *all* moderately suitable land for rainfed *milpa* agriculture with *two* crops per year is taken into account for estimating yields in the study area (64.6 km^2), it is calculated that 6.6 km^2 (S2ic, with gray color, Figure 5.9) would produce 792,000 kg each crop cycle (1,584,000 each year). This maize quantity could feed 5637 people in a good year. In the same way, if *all* moderately suitable land for rainfed *milpa* agriculture with *one* crop per year is taken into account for estimating yields in the study area (64.6 km^2), it is calculated that 28.01 km^2 (S2si, with gray color, Figure 5.10) would produce 3,361,200 kg each year. This maize quantity would feed 11,961 individuals.

These maximum estimations are calculated with first year *milpa*, with all moderately suitable terrain being cultivated at the same time. This condition is hypothetical, so numbers would decrease considering fallow periods. Presently, there is not enough information to determine the duration of the fallow periods. More research related to settlement and demography could afford a more accurate estimation of this variable.

However, it is interesting to note that marginally suitable terrain (*akalche*) could give another panorama. If this terrain could be managed in order to decrease inundation, it could support agriculture and hence more people.

For this reason, hydraulic modifications for rainfed *milpa* with either one or two crops per year should be expected since dry periods significantly threaten agricultural productivity. Nearby meteorological stations with long-term registers placed at Noh Yaxche (24 years) and Laguna Alvarado (5 years) show considerable rain variability (location shown in Figure 3.1). The range of annual precipitation varies from 800 to 1600 mm, with wide fluctuations from year to year (Figure 5.11).

After examining the landscape of Calakmul from the perspective of land evaluation, it is proposed that the land is moderately and marginally suitable for *milpa* system agriculture. This system, combined with horticulture and forest management, could support a small number of inhabitants. However, if this practice was enhanced by terrain management, suitable areas for *milpa* agriculture augmented, and hence, the population increased significantly.

CHAPTER 6. THE ANCIENT AGRICULTURAL LANDSCAPE OF CALAKMUL

"The true secret of Maya civilization was their successful, sustainable adaptation to the rain forest, a feat we have not been able to duplicate today."

Demarest, 2006:128.

Climate and Landscape in the Maya Area

Since climate is intimately linked with agriculture, a general panorama of climatic conditions is presented before addressing issues related to subsistence. The focus of discussion here is how the climate has changed during the last four millennia in the Maya area. Researchers have shown a great deal of interest in climatic history due to the fact that drastic changes in the climate have often been posited as the main factors leading to the Maya Collapse. In one study at Lake Punta Laguna, Quintana Roo, Mexico, Curtis and colleagues discuss the oxygen isotope records ($\delta^{18}O$) of ostracods and gastropods near the archaeological site of Coba (Curtis et al., 1996). Isotopic measurements of both microfossils yielded similar reconstructions of climate. Their results show that Late Holocene climate in the region was marked by repeated wet and dry episodes, and the periods of drought (equivalent to A.D. 862, 986 and 1051, calibrated) correspond to the period of the Maya Collapse.

Likewise, the analysis of the bulk titanium content of undisturbed sediments in the Cariaco Basin in the southern Caribbean Sea reflects an extended regional dry period punctuated at approximately A.D. 810, 860 and 910. It is suggested that these abrupt drought events contributed to the social stresses that led to the Maya demise (Haug et al., 2003).

Along similar lines, Leyden and colleagues undertook a project to reconstruct Holocene environments on Yucatan Peninsula (Leyden et al., 1996). This investigation recorded inferred climatic change in the Peninsula from four lake-sediment stratigraphies (Whitmore et al., 1996). Based on an analysis of sediment cores in San José Chulchaca, Yucatan, they concluded that from 6150 B.C. until the Early Preclassic, the region experienced significantly wetter conditions than are common today, and that the forests of southern Yucatan state may have been transitional in relation to the more humid forests of southern Campeche. Then, a dry period commenced around 1800 B.C., which promoted more open forests. During the last years of the Preclassic leading up to the Early Postclassic, slightly wetter conditions prevailed, followed again by a consecutive dry period (Leyden et al., 1996).

Another study in Lake Coba, Quintana Roo, showed that before 1650 B.C., the indigenous vegetation was approximately half semi-deciduous and half swamp forest. After this date, the inhabitants cleared the land. The investigation also found evidence of maize cultivation around 850 B.C., with lakeside *milpas* that existed until A.D. 720. During the Maya Collapse, Coba was sparsely inhabited, and after A.D. 1240 it was completely abandoned, although some vestiges of its former glories remained (Leyden et al., 1998).

In addition, a palaeo-environmental study from Lago Puerto Arturo, Petén, Guatemala revealed a 3600 years record of human activity in the Mirador Basin from approximately 2700 B.C. to A.D. 900. Pollen analyses reflected an abrupt increase in anthropogenic disturbance during the Early Preclassic. The record indicates at least four phases of agricultural disturbance, with intervening periods of ecological recovery, during the following 2500 years. The last agricultural phase ended in about A.D. 900, coincident with the Late Classic abandonment of the southern Maya lowlands (Wahl, 2005).

Moreover, geographic studies affirm that the Maya might originally have settled in an environmental landscape quite different from the modern one. Dunning and colleagues (Dunning et al., 2002) have proposed that in the Maya Lowlands human-induced environmental change, in tandem with natural climatic shifts, transformed at least some of the karstic depressions, the *bajos*, between 400 B.C and 250 AD. The *bajos* were anthropogenically transformed from perennial wetlands and shallow lakes to seasonal swamps. They assert that these changes may help explain why many of the earliest Maya cities were built around *bajos*, why some early Maya urban centers were abandoned near the end of the Late Preclassic period, and why other centers developed elaborate water-storage systems.

Similarly, Hansen and colleagues suggest that large-scale demographic densities in the Mirador Basin were the consequence of humid, perennially wet marshlands that provided numerous resources of great strategic importance for the major urban centers of the Northern Peten (Hansen et al., 2002:278). Through an interpretation of palaeo-ecological data they suggest that "the Maya Lowlands suffered a transition from a mature lacustrine system consisting of wetland marshes to the forest system that we observe at present" (Hansen, et al., 2002: 280). Hansen and colleagues found a buried A horizon (paleosoil) in Nakbe that is believed to be the original surface of the *bajos* when they were a wetland marsh. The overlying intrusive sediments were associated with a period of intense sedimentation followed by an abandonment of the area as part of the succession to contemporary forest species (Hansen et al., 2002:281).

- Climate and Landscape in Calakmul

Palaeo-environmental studies suggest that two specific periods denoted dry climate conditions for thriving Lowland Maya population: the transition between the Preclassic and Classic periods (Haug et al., 2003, Wahl et al., 2006) and the end of the Classic period (Curtis et al., 1996, Haug et al., 2003, Leyden et al., 1996, Wahl et al., 2006). These dry periods have been identified as the major cause for population decrease and city abandonment. Definitely, the climate had a significant impact in Maya population although other factors should be considered as for example changing social institutions or wars.

Without doubt, the ancient inhabitants of Calakmul had to face an instable climate. Even today, the records of meteorological stations near the Biosphere Reserve of Calakmul show a pattern of variable precipitation (Figure 5.11). This pattern shows a year of abundant rains followed by two or three years of lower precipitation. The records of the Calakmul meteorological station presumably show this same tendency, but since the registers began in 2003 it cannot be confirmed yet. This rain variability reflects the evident necessity of keeping water for the dry periods for maintaining large populations.

Moreover, the two dry periods affected Maya people with a different impact. The drought conditions that suffered Maya inhabitants during the transition between Preclassic and Classic periods did not promote the abandonment of Calakmul, situation that contrast with Peten sites that were deserted such as El Mirador (Webster, 2002). However, the circumstances were different during the time between Classic and Postclassic periods, when Calakmul was abandoned as many other Lowland Maya sites.

Agriculture

Organized agriculture, that is, the continuous cultivation of the land in order to harvest subsistence crops, was a revolutionary activity since it completely changed the ways in which humans related to each other and to their environment. A sedentary lifestyle, population increment, the domestication of plants and animals, and the development of complex social organization are just some factors closely linked to this seminal activity. In this section, the origins of agriculture and the processes of its development in the Maya area are discussed.

During the Pleistocene epoch, humans followed a nomadic way of life, subsisting through fishing, hunting, gathering and processing food. These basic subsistence activities began to be phased out during the Holocene transition, about 10 000 years ago, when shifting ecological conditions proved favorable for agricultural innovations. As a result of drastic environmental transformations, new niches became suitable for human inhabitation. The invention of agriculture increased the number of options available in terms of meeting basic nutritional needs, promoted a sedentary lifestyle, and fundamentally altered the manners in which humans related to their physical environment. The first agriculturists probably began to observe plant cycles, and thus to understand the influence of factors such as the elements, soil, or insects, and likely initiated the domesticated cultivation of some wild species. It has been suggested that during the Holocene, a rise in atmospheric CO^2 levels increased plant productivity by up to fifty percent, a phenomenon that would have greatly boosted the viability of sustained, sedentary agriculture (Sage, 1995).

Although the transition processes that led from hunter-gatherer groups to agricultural communities are still the subject of considerable debate, human agricultural knowledge and practices eventually and subsequently led to the selection of sites favorable to the activity, recognition of the optimal candidates for cultivation, and the invention of storage alternatives and food processing techniques. Gradually, these changes encouraged a distinct way of life near those places most suitable for supporting the long-term procurement and harvesting of resources.

Agriculture appeared in distinct spatial-temporal contexts, in different regions that correspond to present-day areas in the Near East, Ethiopia, Sub-Saharan Africa, China, highland and lowland South America and Mesoamerica (Harlan, 1997). Crops such as rice, maize, wheat, millet, potato or sorghum, among others, were transformed into the primary staples for the world's early villagers. In Central America early agricultural activity is inferred from the starch grains of manioc, yams and arrowroot, dated from between 7000 to 5000 years ago (Piperno et al., 2000). These crop findings have led researchers to propose that agriculture originated in the humid, tropical settlements (Pope et al., 2001).

Agriculture in Mesoamerica

In Mesoamerica, forty-nine species were domesticated including maize, beans, squash, spices, fruits, and fiber plants (Hawkes, 1998). The first domesticated species was squash, cultivated in Oaxaca from 8000 to 10 000 years ago (Smith, 1997). More than three thousand years later maize began to be cultivated.

Evidence of maize in the archaeological record dates to 5400 B.C. in Oaxaca, and to 5000 B.C. in San Andrés, Tabasco (Piperno and Flannery, 2001; Pope, et al. 2001). Maize cobs from the San Marcos Cave in the Valley of Tehuacán have been dated to 4,700 14C years B.P. (Long et al., 1989). The presence of maize from the Archaic Period has also been reported in Zohapilco (within the Basin of Mexico), in the Oaxaca Valley, central and southern Tamaulipas, in the Santa Marta Cave in Chiapas, and in Yohoa, Honduras (McClung, 1997). Fray Diego de Landa noted that Maya had very good maize, in a multitude of varieties and colors, which was stored in silos for consumption in lean years (de Landa, 1994). Although the landscape varied within the Maya area, the basic diet was shared between almost all groups.

Other important crops were available to the Maya. As it is written in Popol Vuh:

"And in this way they were filled with joy, because they had found a beautiful land, full of pleasures, abundant in ears of yellow corn and ears of white corn, and abundant also in *pataxte* and cacao, and in innumerable *zapotes*, *anonas*, *jocotes*, *nantzes*, *matasanos*, and honey. There was an abundance of delicious food in those villages called Paxil and Cayalá. There were foods of every kind, small and large foods, small plants and large plants" (Recinos et al., 1954:105).

Beans (*Phaseolus vulgaris*) comprised another primary, essential crop. In fact, beans and maize constitute a nutritionally complete meal. Squash and gourds, whose flowers and seeds were also consumed, and which, when dried, were even used as containers, were another common cultigen. Re-analyzed *Cucurbita pepo* seeds and fragments dating to between 8000 and 10 000 years ago, were found in Guila Naquitz cave in Oaxaca (Smith, 1997). Additionally, lima bean (*Phaseolus lunatus*), avocado (*Persea americana*), chili pepper (*Capsicum annum* and *Capsicum frutescens)*, manioc (*Manihot esculenta*), agave (*Agave frourcroydes*), cotton, sunflower, rootcrop (*Pachyrrhizus erosus*), chaya (*Cnidoscolus acotinifolius*), pepper (*Pimenta dioica*), vanilla (*Vanilla planifolia*), guava (*Psidium guajava*), hug plum (*Spondias purpurea*), papaya (*Carica papaya*), black zapote (*Diospyros digyna*), nance (*Byrsonima crassifolia*), pitahaya (*Hylocereus undatus*) and cacao (*Theobroma cacao*) were valuable plants that were also cultivated (Colunga and Zizumbo, 2004; Pope et al., 2001).

Around 2500 B.C. cotton and the sunflower cultivation appear to have emerged in Tabasco area (Pope et al., 2001). Little is known about the cultivation of cotton, but Fray Diego de Landa wrote that it grew everywhere in Yucatan, in two varieties: an annual plant and a species that lasts for five or six years (de Landa, 1994).

Although evidence of manioc cultivation has been found in the Maya area dating from about 4600 B.C. (Pope et al., 2001), the role of root crops has not yet been fully comprehended. De Landa mentions five edible roots (two of them consumed in periods of famine; de Landa, 1994). Hellmuth found sixteenth and seventeenth century documents that reported that the Maya ate *camote* (sweet potato), *jícama* (rootcrop), *macal* or *yautia* (*Xantosoma yucatense*) and yucca (Hellmuth, 1977). However, since evidence of these crops is rarely found in excavations (probably because root crop remains do not preserve well in tropical environment), little is known about the specific sowing and harvesting practices associated with them.

One cannot speak of agriculture without referring to plant domestication. Domestication is the genetic alteration of certain plants that results out of the intentional human selection of desired qualities, such as bigger size, better flavour, particular use, etc. (McClung, 1997). Consequently, domesticated species became dependent on humans. Early agriculturalists must have learned to save the seed and modify the ground in order to favor growing conditions. Based on macro-remains, Lenz proposes that by 1200 B.C. the Maya had a solid maize-based system for food production. Other important species that formed part of this system were squash, chili peppers and different kinds of beans. With similar evidence, he also proposes that perennial tree crops were combined with the production of annuals to satisfy the demands of an increasing population (Lenz, 1999).

It has been shown that agriculture and animal domestication were closely interrelated. About 8000 years ago the dog was domesticated in North America and approximately 2000 years later, the turkey in Mesoamerica (Valadez, 1999). Both turkey and dog were raised for human consumption in Mesoamerica. Fray Diego de Landa reports that the Maya ate stewed vegetables, deer meat, wild and domestic birds and many varieties of fish (de Landa, 1994). There is no evidence for the existence of beasts of burden, nor the use of animals in plowing. Consequently, the primary agricultural workforce was comprised of human beings.

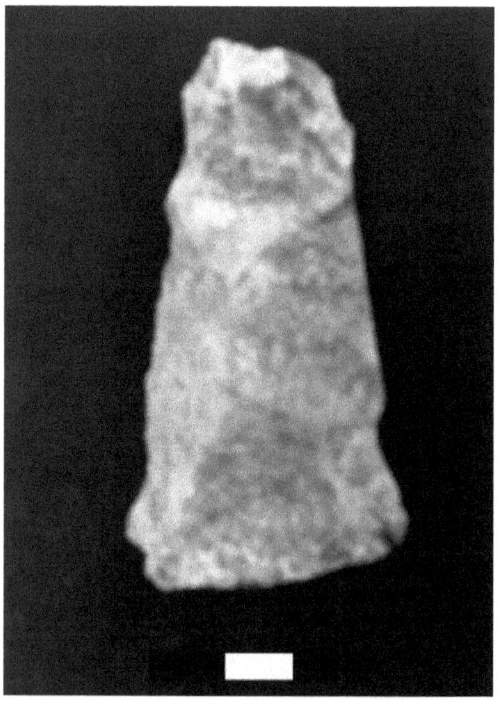

Figure 6 1. Limestone pick prepared for hafting in a wooden handle. It was found in Calakmul in Pit J. The scale measures 3 cm.

Metal tools were not available to assist in land clearing cultivation in the Maya area, at least until the Postclassic period. The principal agricultural tools were made up of flint or limestone with wooden handles (See Figure 6.1). A baton-like stick known as a *coa* was used for sowing. Grinding stones were indispensable for processing cereals. These were of limestone or imported material from the southern highlands. And although currently difficult to identify accurately, storage places were probably built to safeguard food for lean periods.

Some options for fertilization and soil preparation in prehispanic times were the use of partially decomposed aquatic plants, bat and turkey guano, human faeces, food scraps, ashes, burned bones, and other composted, organic materials (Fedick and Morrison, 2004). Additionally, the use of water from ancient irrigation canals (with their high content of organic matter) and practices such as crop rotation would have ensured better yields (Flores Díaz, 1975). For example, in the Maya lowlands, the use of periphyton (a complex algae, and other organisms that fix phosphorous and nitrogen in wetland soils) could have enhanced and improved such terrain for agricultural purposes (Morrison and Cózatl Manzano, 2003).

Regarding pest control, scientists have identified twenty-two species that pose a threat to *milpas* (intercropped maize, bean and squash plots of land) in Central America (CATIE, 1990). Modern Guatemalan farmers affirm that pests do not attack *milpas* because they know when to prepare the land, when to sow and harvest. They say that the weather and soil are favorable to the cultivation of corn and insects are almost non-existent if they cultivate following the tradition of the elders. Moreover, they practice twenty-six preventive actions to repel insects such as respecting planting dates, sowing and harvesting on the full moon, actively protecting and nurturing the natural enemies of unwanted insects, or inter-planting corn with beans to avoid beetles (Morales and Perfecto, 2000).

In the Maya area, a profound knowledge of the climate, astronomy, and the environment allowed the inhabitants to develop a complex agricultural system, which allowed demographic increase. In order to sustain such a large population the Maya necessarily had fixed agricultural practices. They understood the relationships between the landscape features and adapted their practices to maintain their basic needs in terms of soil fertility, the availability of water and habitat diversity.

Ancient Maya Agriculture

Research in Belize, through the identification of fossilized pollen, confirms that agriculture was present there before 3000 B.C. (with the introduction of maize and manioc), and that in 2500 B.C. dramatic deforestation in the Maya lowlands was evident (Pohl, 1990). Proxy records (natural sources of climate information such as ice cores, lake and ocean sediments or pollen) from Lake Salpeten in Central Peten, indicate that the Maya began to clear the forest and consequently induced soil erosion beginning around 1700 B.C. Many centuries later, by A.D. 850, soil erosion was markedly reduced, a trend that coincided with the Terminal Classic Maya demographic decline. Following this, with further depopulation of the region, the forest recovered and organic carbon sedimentation increased after A.D. 1400 (Rosenmeier et al., 2002).

Nearby, a long core from Lake Peten Itza revealed deforestation promoted by prehistoric Maya inhabitants. This is shown by a pollen record that began about 2000 B.P. with the appearance of disturbance taxa (e.g. *Ambrosia* [ragweed] and Poacea [grain cereal family]). After the Classic Maya collapse, forest re-growth occurred as was reflected by a relative increase in *Moraceae-Urticaceae* pollen (trees as *Brosimum* or *Ficus* and weed; Curtis et al., 1998; Islebe et al., 1996). In concordance with these data and partially correlating with deforestation episodes, through a paleohydrological model, Gunn and colleagues propose that southern lowland Maya thrived when the environment posed an optimal balance of wet and dry season duration; in contrast, they fell down during extended wet or dry periods (Gunn et al., 1995).

- Myths Concerning Ancient Maya Agriculture

There have been some myths that have influenced the way we conceive Maya agriculture. After years of research many of them seem obvious but, a brief recapitulation is nevertheless useful to emphasize their importance. The first one is that contemporary tropical agro-ecosystems are neither stable nor especially productive (Sanchez and Logan, 1992). Factors such as low yields related to increased poverty levels and government corruption have augmented the apparent veracity of this assessment. The "globalized" world market economy has forced farmers to cultivate in a non-sustainable way, using non-organic pesticides and fertilizers that in the long term deplete the land. Further, the thickness of tropical soil, as well as the presence of extended drought periods and pests, serves to enhance the perception of a restrictive instability. Although this situation might be true in many modern tropical countries in Central America, the panorama would have been different in the past. The ancient Maya survived in similar conditions for millennia, with constant climatic variation, but still they were able to cope with their environment and sustain a remarkable civilization.

Stemming from this assumption is the notion that tropical terrain is simply not adequate for sustained, intensive agriculture (Sanchez and Logan, 1992). This supposition may be accurate if we only were to consider "sustained, intensive agriculture" for "green revolution" (modern, industrial) agriculture, for which tropical land is simply not apt. "Green revolution" agriculture involves the introduction of high-yield varieties of grains, mechanized management and irrigation, and the use of non-organic fertilizers as well as pesticides. These practices have been shown over time to be unsustainable, and exhaust the terrain relatively quickly, eroding the land and converting rainforests into unproductive wasteland for decades.

Another fact that contributes to the perception of Calakmul's environment as inhospitable to human habitation is that recent colonizers have had great trouble in settling the Maya rainforest (García Gil and Pat, 2000). Water scarcity is the main factor that impedes steady development. There are many small communities that have been founded in the past few decades that still lack basic services and organization (Morales and Magaña, 2001). These people come from other parts of the Mexican Republic and do often have no experience in coping with or managing tropical environments (Gurri, 2006). Hence, their adaptive process is one of trial and error, often with negative results, which culminates in the abandonment of the land. In addition to lack of environmental knowledge, the deficiency of social organization minimizes the possibility of successfully populating the region.

Finally, the assumption that the swidden system (slash and burn method of terrain preparation for cultivation) is very destructive and its yields are marginal has been thoroughly discussed (Cowgill, 1961; Pool, 2001). Maya swidden agriculturalists carried out practices that avoided exhausting soils and in turn allowed for fallow periods and subsequent regeneration (Barrera et al., 1977). This system had been adapted to the conditions of each site, and combined with other subsistence options may have been necessary in order to sustain a large population.

- The *Milpa*

Mesoamerican populations widely practiced an agriculture system with maize as the main crop. This system refers to a maize fallow-based cultivation employing slash-and-burn techniques. Significant hints regarding Maya slash-and-burn agriculture are found in the Popol Vuh:

"Soon they [Hunahpú and Xbalanqué] came to the field. And as they plunged the pick into the earth, it worked the earth; it did the work alone. In the same way they put the ax in the trunks of the trees and in the branches, and instantly they fell and all the trees and vines were lying on the ground. The trees fell quickly, with only one stroke of the ax. The pick also dug a great deal. One could not count the thistles and brambles which had been felled with one blow of the pick. Neither was it possible to tell what it had dug and broken up, in all the large and small woods." (Recinos et al., 1954:65).

The traditional Mesoamerican *milpa* system, called *col* in Maya, consists of an intercropped cultivation of maize (*Zea mays*), beans (Phaseolus spp) and squash (Curcubita spp) with variable fallow periods and the use of mulch (See Figure 6.2). Through the slash and burn technique, Maya farmers clear an area (of variable size) of excess vegetation and then sow maize, bean and squash seeds. After harvesting, they use the terrain for one or two more years, or let it rest until the native vegetation re-grows, the so-called fallow period, and repeat the slash and burn practices in a different parcel. This process is observable in hundreds of present-day farms in the Yucatan Peninsula, although this type of practice is diminishing because former farmers are being employed outside their homes in larger cities or other countries.

Figure 6.2. Milpa system. Notice the intercropping of maize, beans and squash.

The primary question here is if ancient and modern Maya agriculturalists employed a similar agricultural system. According to Terán and Rasmussen, the contemporary and prehispanic *milpa* systems are quite alike, but today's system is just a pale reflection of the ancient one (Terán and Rasmussen, 1995). The main difference is that current fallow periods are shorter. The similarities have been preserved due to the nutritive value of maize and its cultural importance, since it is a crop that is intimately related to the Maya worldview and social organization.

Accounts related to and describing agricultural systems can be encountered in Colonial period literature. During the sixteenth century, a simplified form of the long fallow, swidden agriculture was widely observed by the Spanish conquerors (Dunning and Beach, 2000). Consequently, the Spaniards identified this system as typical of the region. But given that social circumstances had been drastically altered since the Classic period through the conquest, probably this observation was not completely accurate (Dunning and Beach, 2000).

The key to the success of the *milpa* system over time is its diversity. Although the main species, maize, squash and beans are intercropped in the parcel, some sections are dedicated to other cultigens such as tubers, chilies, fruits and herbs. Moreover, short and long-term species are cultivated according to specific cycles, and other activities are practiced around the *milpa*, such as collecting wood (Terán and Rasmussen, 1995). In addition, some useful species are protected from elimination during the slash and burn clearance process.

The first step for cultivating a *milpa* is to clear the terrain, primarily through the slash and burn technique. This generally occurs the months of March or April (See Table 6.1). The bush is felled, and smaller trees and plants are cut up, piled and allowed to dry. Late in April or May the bush is burned. Some valuable species, such as *zapote*, *ramon* or thatch palms, are sometimes spared (Lundell, 1937). Through the fire, weeds and insects are killed, and the ashes are converted into a nutrient for the subsequent crop. Then the earth is left undisturbed until the advent of the rainy season in May or June, when farmers begin to sow. With the aid of a sharp-pointed stick, they make a 10 cm hole and plant seeds of maize, bean and squash. At the same time many weeds grow, so in July farmers must hand-weed the area in order to avoid competition for the soil, water and solar resources. The maize grows until September or October, at which point the stalks are bent at the nodes below the ears of maize. This action is useful for drying the cob, protecting it from birds and preserving the maize until it is finally harvested for use. Maize can even be harvested as late as February, if need be (Arias, 1994; Graefe, 2003; Lundell, 1937).

In an interesting study that compares the traditional Maya with modern agricultural methods (that include the use of non-organic fertilizers and insecticides), Graefe studied the environmental factors related to the *milpa* system. The traditional farmers whom she investigated live in the towns of Becanchen and Progresito, located in the south of the state of Yucatan. She found that these traditional farmers employ a fallow period of up to 14 to 50 years and their total *milpa* cultivation area for each year was between 4.4 and 6 ha (Graefe, 2003:54). She also observed that farmers plant different maize varieties in their plots, which can be observed in Tables 6.2. and 6.3. Even in the Popol Vuh this practice can be noticed: "From Paxil, from Cayalá [mythological places], as they were called, came the yellow ears of corn and the white ears of corn (Recinos et al., 1954:104).

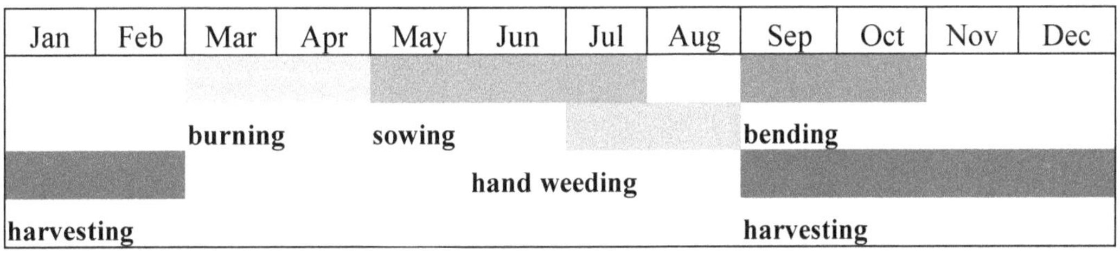

Table 6.1. The milpa system calendar (from Arias, 1994 and four traditional farmers from Yucatan personal communication, 2005).

Maya Farmer	*Milpa roza* varieties	*Milpa caña* varieties	*Milpa caña k'ankab* soil varieties	*Solar* (garden) varieties
Valerio	*xnuc-nal*	*xnuc-nal*	*nal xoy*	no
			tsiit-bakal	
Ignacio	*xnuc-nal*	*xnuc-nal*	no	*nal-tel*
	white variety	*tsiit-bakal*		
	xhe-ub			

Table 6.2. Milpa varieties sown in different types of terrain (based on Graefe, 2003:57). In milpa roza, the plants were sown in land recently cleared by burning; in milpa caña, plants were already cultivated in the previous year.

Maya farmer	Variety	*Milpa roza* Yield (kg/ ha)	Variety	*Milpa caña* Yield kg/ ha)
Ignacio	*xnuc-nal*	1250	*tsiit-bakal*	1500
	white maize	875	*nal-tel*	1375
	xhe-ub	1375		
Valerio	*xnuc-nal*	800-1000	*xnuc-nal*	500-600
			nal-xoy	1500-1750

Table 6.3. Yield of maize varieties for traditional Maya farmers, year 2000 (based on Graefe, 2003:36,61).

These farmers own terrain with *tsek'el, k'ankab* and *kakab* (the characteristics of these soils have been described in Chapter 6). They sow the *xnuc-nal* variety in the *milpa roza* (recently cleared by burning) and *milpa caña* (cultivated in the previous year) located in *tsek'el* terrain. They use other varieties, as is evident from the table above, for the other kinds of soils. One of them also sows maize *nal-tel* in his *solar* (garden). The yield of each kind of maize also varies a lot as seen in Table 6.3.

The contemporary farmers who live around Calakmul also cultivate maize as a subsistence crop within the *milpa* system. They plant corn, beans and squash together for two or three years and then allow for a fallow period of five to twenty years, depending on the location and characteristics of the terrain (Linteau, 1996). In the Municipio of Calakmul, the land is held under the *ejido* system (an *ejido* is a large tract of land that is communally owned and shared by a group of farmers, *ejidatarios*). Each farmer has a specific allotment of land for cultivation, and a swath of rainforest is usually shared between the members.

The *ejidatarios* of Calakmul farm using the traditional *milpa* systems, and have also cultivated jalapeño peppers as an important cash crop since the 1980s. Their yields are very low and their options for timber harvesting are restricted, since the most valuable species have been previously harvested. Morales and Magaña report that traditional *milpas* in the Calakmul area attain yields of 0.5 ton/ha (i.e., 500 kg/ha; Morales and Magaña, 2001). However, limited forest exploitation is still a source of income for many communities, as are cattle ranching, apiculture, and the collection of *chicle* latex (Linteau, 1996).

Estimates from diferent regions of the Yucatan peninsula are useful for giving an idea of the average yield of a Maya farmer. The yield calculations from Peten, northern and southern Yucatan are shown in Table 6.4, except for the one of Murtha that is based in an average number obtained from "hundreds of studies" (Murtha, 2002:217), and that he used for demographic estimates of the prehispanic site of Caracol, Belize. In this table it is also possible to observe that the first year crop yields an average of 1173.56 kg/hectare, and the second, 861.28 kg/hectare (about 25% less).

Preparation of the terrain is another decisive factor for yield variability. In an experimental *milpa* parcel in Yaxcaba, Yucatan, scholars avoided burning the forest vegetation before sowing. As a result, more man-hours of work were required to preserve the *milpa* and the yield decreased (Pool, 2001).

The farming of maize as a basic crop gives us a general idea of currently and previously viable subsistence possibilities. However, other cultigens were, and remain, of the utmost importance. Maize and beans are often the only cultigens commonly contemplated as part of archaeological yield calculations, although other crops, such as squash or chili, are worthy of consideration. In Table 6.5, the average density of the major *milpa* plants is shown. This table is included in order to give an idea of the number of plants in a given area. As the data show, the density evinces a great variability, which will be also reflected in the yield. These data are relevant in terms of illustrating the difficulty of archaeologically calculating the yields of crops due to the complexity of the variables involved.

In addition to *milpas*, modern Maya cultivate other crops in plots near their houses. These plots have perennial herbs tubers as *macal* or sweet potatoes, as well as peppers or tomatoes (Lundell, 1937). Other secondary cultivated areas have trees such as avocado, *caimito*, guava or cacao. Maya farmers also have a garden in their household. In this place they usually cultivate spices, fruits, medicinal plants, flowers and ornamental plants.

First year kg/ha	Second year kg/ha	Site	Reference
1200.00	no	General estimate	Murtha, 2002:217
1597.42	1132.00	Peten Itza	Cowgill, 1962:276
1042.53	874.38	Northern Yucatan	Steggerda, 1941 in Cowgill, 1962:277
1375.00	no	Chichen Itza	Morley, 1947:154
850.00	375.00	Chichen Itza	Morley, 1947:153
1250.00	1375.00	Southern Yucatan	Graefe, 2003:36,61
900.00	550.00	Southern Yucatan	Graefe, 2003:36,61
1173.56	861.28	--	AVERAGE

Table 6.4. Yield of maize varieties for first and second year crops for Maya farmers.

Likewise, the ancient Maya cultivated gardens which incorporated valuable crops, a fact born out by the available archaeological evidence. Spaces between residential units were used for gardening (Drennan, 1988), places with fertile soil were obviously preferred, (Fedick, 1995) and fertilizer was used to improve soil and harvests (Fedick and Morrison, 2004). Furthermore, the *pet cot*, an enclosed vegetation area surrounded by a stone wall has been documented to contain a many as twenty-nine valuable species like ramon, anona or elemuy. The *pet cot* appears to be a remnant of ancient agricultural practices of Yucatan Peninsula (Gómez Pompa et al., 1987).

Crop	Average density (plants/ ha^{-1})	Minimum and maximum (density plants/ ha^{-1})
Maize	25 000	20 000-30 000
Bean	1800	400-4 270
Squash	1900	930-4 900

Table 6.5. Average density of a traditional milpa in the southern Yucatan peninsula (based on Graefe, 2003:59).

Wetland Agriculture

Wetland agriculture is related to cultivating in lands that are inundated or saturated by surface or groundwater at a frequency and duration sufficient to support a prevalence of hydrophytic vegetation. A wetland area is inundated for at least seven days or saturated for fourteen consecutive days during the growing season. Native wetland vegetation is adapted for life in saturated soil conditions (http://www.cbbel-in.com/cbbelwetland/ Definitions.htm, accessed August 22, 2004). Some areas in Calakmul, specifically karstic depressions, satisfy these conditions and can be considered areas for wetland agriculture.

Although wetland regions may appear to be marginal areas for agriculture, they do support diverse fauna and can be cultivated. Their vegetation is a source of fibers and wood. The use of wetlands greatly increased the space available to the Maya for the practice of agriculture. Evidence for wetland agriculture has been found for example in Belize, at San Antonio Rio Hondo. Excavations showed that the swampy terrain was cultivated with dry-season cropping, with ditches constructed to encourage drainage (Bloom et al., 1983). Wetland cultivation has also been explored in seeking to explain how ancient Maya settlements supported dense populations (Culbert, 1995). Moreover, the use and exploitation of *bajos* has been a matter of heated debate (Culbert et al., 1990, Pope and Dahlin, 1989).

From 1500 to 1000 B.C., wetland agriculture was introduced in northern Belize. In 1000 B.C, some hydraulic constructions were built to compensate for the rise in groundwater levels (Pohl et al., 1996). Investigations in northwestern Belize at La Milpa *bajo* reveal that the small Preclassic population practiced extensive forms of forest clearance and agriculture that completely modified the landscape. This alteration was reflected as a change from perennial to seasonal wetlands.

As a consequence, subsequent generations were left with an eroded setting in later periods (Dunning et al., 1999).

The Maya adopted different wetland agricultural practices for managing terrain such as making terraces (Fedick 1995; Healey et al., 1983; Murtha, 2002) or raised fields (Puleston, 1978; Siemens and Puleston, 1972; Turner and Harrison, 1981; 1983). These practices evidence diversified and intensive agriculture. Intensive agriculture refers to agricultural practices that produce high output per unit area usually by intensive use of labor (United Nations, 1997). Other practices such as shortening fallow periods, cropping more than once a year or introducing new cultigens also increase the output per unit area.

Agricultural terracing provides direct evidence of human adaptation of the land to sustain production (Beach et al., 2002). The construction of terraces was a result of the intensification of production in response to a rapidly growing population, and the use of terraces likely increased during the Late Classic period. These features were built in land with variable degrees of inclination, in order to divert and slow runoff, and to build up planting surfaces that could maximize soil moisture (Beach et al., 2002). For example, in the Upper Belize River area terraces were located in terrains with low slope within soils developing on consolidated limestone. They are commonly associated with densely settled uplands that possessed better quality natural resources conducive to increased agricultural capability (Fedick, 1994). Terraces have also been reported for example in Caracol (Murtha, 2002), in the Three Rivers region of Belize (Beach et al., 2002) and in Yaxchilan (Lobato, 1988).

The building of wetland fields (drained or raised fields) represents another option for taking advantage of the landscape. Since they were detected in the 1970s (Siemens and Puleston, 1972) many other researchers have reported them (Adams et al., 1981; Bloom et al., 1983; Pohl, 1990; Pope and Dahlin, 1989; Turner, 1983; Turner and Harrison, 1983). Drained fields are present in land surfaces with canals for mobilizing water. Raised fields are artificial elevated plots divided by channels built with soil from elsewhere.

Wetland fields in Maya area avoid inundation of flat areas, as *bajos*. They increase the available land for agriculture, hold nutrients and retain humidity necessary for plant growth. These features were probably built with the aim of satisfying local agricultural production, but only a subset on specific localities (of hundreds of thousands of natural wetlands in the Maya lowlands), may have been suitable for cultivation (Fedick, 1996). Raised fields have been considered an early adaptation of Preclassic period lowland people (Pohl, et al., 1996), but they were also used during the peak of the Classic period (Pyburn, 1996).

- Wetland Agriculture at Calakmul

Land features that resemble a wetland field system were observed at Calakmul (see Figures 3.9 to 3.13). After the field reconnaissance and aerial photographs observation, their presence in the northern and eastern section of Calakmul was suggested. First, the big quantity of streams detected during the exploration of the site during 2003 and 2004 season denoted me this possibility. Second, their localization through aerial photographs and remote sensing images (seen as lattices in some areas) supported this assumption.

It is feasible that in Calakmul *bajos* were farmed, just as in other lowland Maya sites (Beach et al., 2003; Culbert et al., 1990; Sever and Irwin, 2003). At the least, the transitional surfaces between *bajos* and hills provide evidence of this fact. Land evaluation analysis supports also this affirmation. Nevertheless, other researchers disagree with this affirmation (Pohl et al. 1996, Pope and Dahlin, 1989). However, as Dunning (1996) has sagely indicated, the Maya region is very diverse and its investigators should avoid generalizations.

Thus, it is proposed that *bajos* edges were also used for agriculture, particularly for medium-fallow shifting maize and bean agriculture. Calakmul environment satisfies *milpa* climate, temperature and soil requirements. The particular limitations of maize in this zone are water logging and continuous cropping in the same area, unless an organic fertilizer is applied or maize is rotated with leguminous crops (ILACO, 1981). Also, annual variations in rainfall imply a necessary modification of the land with hydraulic works in order to compensate for dry periods (which they did).

Evidence of bajo agriculture is supported by the land evaluation results. The most suitable area for agriculture in Calakmul is that which is comprised of Vertisols associated with Rendzinas, followed by the area whose soil is primarily Rendzinas and Gleysols. Soil distribution maps were presented in Chapter 5. However, a map ranking the agricultural potential of the land can be generated as well, evaluating the area according to slope, soil and vegetation, in order to identify the optimal areas for cultivation (See Figure 6.3). For this purpose, vegetation, soil and slope layers were overlaid in Autocad Map 2002.

It is not surprising that the optimal area for cultivation was located in the transition between the hills and the *bajos* (Figure 6.3, No 1). This terrain has good potential for agriculture because its soils are moderately deep, moderately fertile and irrigated by the *corriental* (and adjoining streams).

The second most viable option for agriculture is the region shown with No. 2 in Figure 6.3. This surface has an elevated percentage of Rendzinas associated with Vertisols. This area has also depressions that canalize water to channels and dolines. Most of the structures of Calakmul are located here. Some settlement studies have shown that the Maya practiced a scattered village lifestyle (Scarborough et al., 1995) and that the space between the constructions was probably used for agriculture (Drennan, 1988). It is likely that this same situation happened at Calakmul. The land concavity between the habitational units could have been useful for catching water and diverting water for agriculture.

Finally, the area shown with No. 3 comprised mainly of Gleysols and the one shown with No. 4, Rendzinas located in slopes. These areas could have supported cultivation, although with many restrictions. The only area that is wholly unsuitable for agriculture is located on very steep slopes, in rock outcrops where Lithosols are found (No. 5).

In relation to the agricultural utilization of *bajos* three hypotheses have been proposed. Essentially, the first proposes that *akalche* areas were lakes during the initial Maya occupation, but after an environmental transition (that transformed lakes into karstic depressions), Maya cities were abandoned (Cook, 1931). The second theory suggests that the *bajos* were historically similar to their current forms (Pope and Dahlin, 1989). Without land management, this would have restricted the agricultural potential of the land, until specific agricultural techniques were practiced. The third theory proposes that *bajos* were tropical wetland marshes that provided resources such as water, wildlife and good soils. These marshlands or *civales* then evolved into *bajos*. Hence, actual *civales* represent fossil remains of ancient marshes (Hansen et al., 2002). The presence of terraces around *akalche* edges, as well as mounds and artifacts, scattered through all but the lower paleosol at *bajos* in Northwestern Belize, supports this theory (Beach et al., 2003). No terraces have been observed at Calakmul, but they were not searched in a systematically way. *Bajos* probably were cultivated in Calakmul, although admittedly paleosols have not been found (yet), nor were *civales*.

Maize, as well as other crops, like cotton, could have been cultivated in the transitional area between hills and *bajos*. In fact, wetlands present optimal conditions for cotton cultivation. In order to grow, this crop needs temperatures of 20° C and 30° C, lots of water, moderate winds, and clayish deep soils that retain water but do not inundate the land for long periods. The best irrigation method is through furrows. http://www.infoagro.com/ herbaceos/ industriales / algodon.asp, accessed July 2005). Although we know that cotton was indispensable for Maya, it has been difficult to detect it in archaeological excavations.

As regards soil management, it has been shown that the Maya manipulated soils in adapting to their physical environment. At present, it is known that some tropical soils are well suited for different crops (Sivakumar et al., 1992). Moreover, human-modified soils can improve the crop production. For example, the construction of terraces has been reported to be restricted to soils of the Mollisol order, located in low slopes developing on consolidated limestone (Fedick, 1994).

Rainforest Management

Researchers suggest that modern rainforest flora composition in the Maya area indicates an integral management of the resources in ancient times. "Poly-cultivation", the selection of plant varieties, the utilization of vertical and horizontal vegetation space, terraces, drained fields, irrigation canals, and even slash-and-burn agriculture were the principal practices for rainforest management (Barrera et al., 1977).

The incorporation of rainforest resources was crucial to Maya subsistence. The Maya managed the forest in order to extract timber for construction, wood for fire, harvest medicinal plants, and obtain food. Among the more significant species is the *ramon* (*Brosimum alicastrum*). In Calakmul, several *ramon* trees in different developmental stages can be observed. Investigators have proposed that the *ramon* seed, eaten raw or as a black flour, could have played an important role in the diet of the ancient Maya (Puleston, 1982). Even today, people make flour from the seeds of the *ramon* as a supplementary food source in periods of scarcity, and also as feed for their domestic animals. It has been suggested that dominant presence of *ramon* at various sites can be traced back to their existence in great numbers at the time these sites were abandoned, so they would likely represent a relict of ancient agricultural practices (Brown, 2005; Lundell, 1937).

Similarly, the *zapote (Manilkara zapota)* is also a prevailing species in Calakmul. The great majority of trees still exhibit several scars, evidence of the collection of the tree's gum-like sap in the mid 20th century. *Zapote* was eaten as a fruit, and its wood was used as for construction and tool making. It has been suggested that its prominence is a consequence of human influence, since during the process of slash-and-burn terrain clearing some of these trees were spared, thus giving this species a survival advantage (Brown, 2005; Lundell, 1937).

Trees valued for their fruits and abundant in Calakmul are sapodilla (*Manilkara zapota*), guaya (*Talisia olivaeformis*), ciricote (*Cordia dodecandra)* and ramon (*Brosimum alicastrum*). Species valued for their wood are chakte vigas (*Sweetia panamensis),* cedar (*Cedrela mexicana*), gumbo limbo (*Bursera simaruba)*, a tree useful for living fences and *tzizilche (Gymnopodium antigonoides)* a tree used as fuel. Finally, two other rainforest plants with significant attributes and cultural usages are copal (*Protium copal*) or *pom*, which produces incense for ceremonies and allspice (*Pimienta dioica*).

Floral and faunal resources were crucial for the subsistence of the population of Calakmul. For example, the analysis of skeletal remains sheds light on how these resources were incorporated directly into human

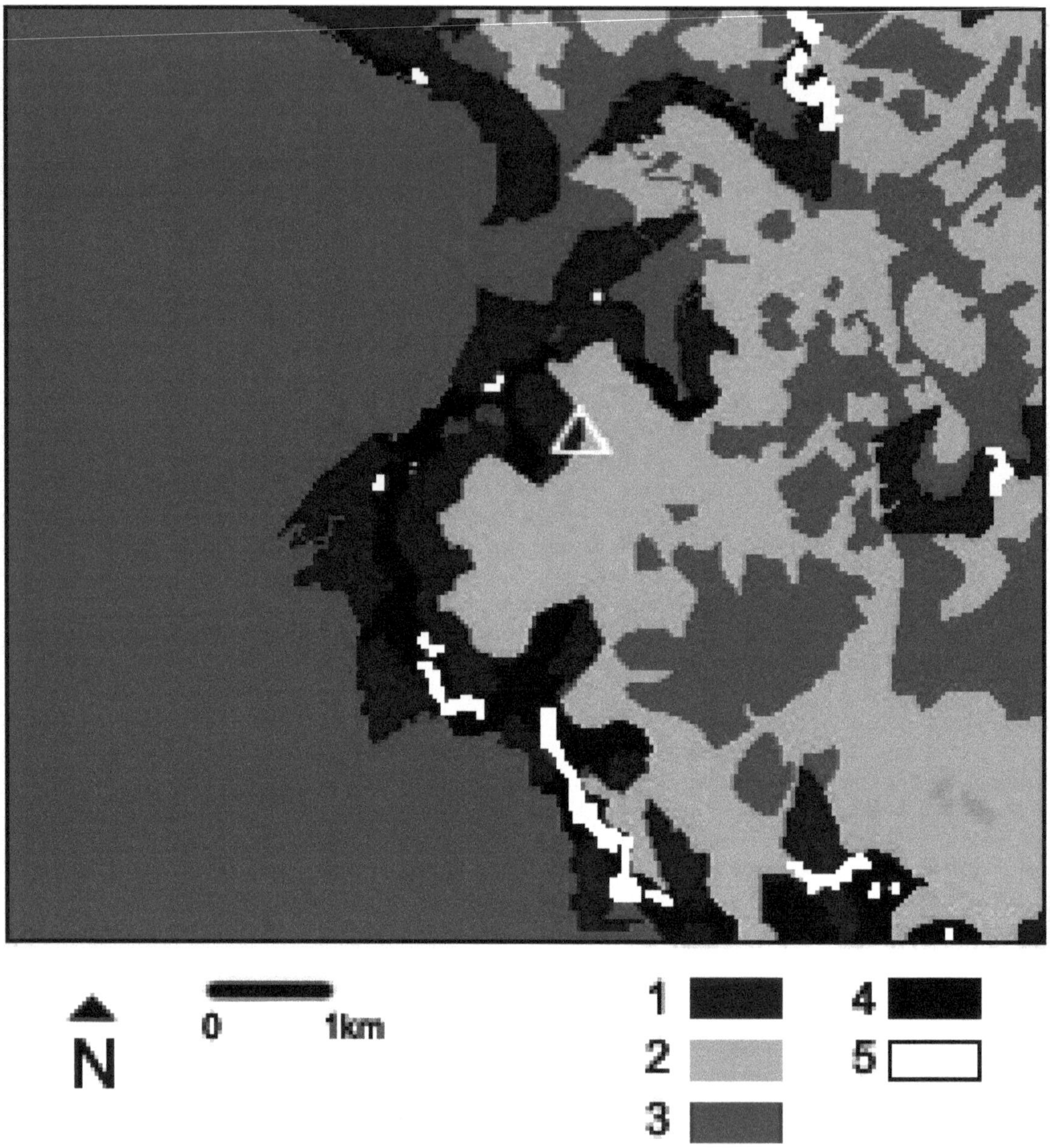

Figure 6.3. Optimal areas for agriculture in Calakmul. Prime land is shown with No. 1. The following numbers show terrain with successively less agricultural potential. The scale measures 1 km; the white triangle indicates the site center.

organism. On the basis of bone analysis, there have been some preliminary attempts to elucidate the basic diet of the population of Calakmul. The main problem is that very few skeletons have been excavated and it has therefore been difficult to carry out a representative analysis of the population. Notwithstanding the great importance of this city and more than twenty years of excavation, we are still unaware of the inhumation practices of the commoners, where their graves are located, if they were even buried (in some particular place), if they were cremated, or if bones are not preserved.

With a heterogeneous skeleton sample (from Calakmul funerary and non-funerary placements) that contained nineteen individuals, males and females, between 15 to 50 years of age and of distinct social status, Tejeda and colleagues (Tejeda et al., 2001) measured the levels of strontium (Sr), iron (Fe) and zinc (Zn) of the bones. High levels of Sr are related to a vegetarian diet, whereas high levels of Zn and Cu are indicative of a carnivorous diet (Tejeda et al., 2001). They found differences between males and females; men appear to have consumed more animal proteins, while women ate more vegetables.

In another study that examined the isotopes of two human skeletons of Calakmul, Wright concludes that dl3c (carbon isotope ratios) values indicate that the consumption of maize in Calakmul was, on average, less than in the northern Peten or eastern Belize. The dl5N (nitrogen isotope ratios) value indicates that ancient people of Calakmul consumed more meat or fish than their counterparts in Peten and Belize (Carrasco et al., 1999a). However, since the analysis is based only in two individuals it is exceedingly difficult to draw generalized conclusions from these results.

- Rain Harvesting in Calakmul Forest

Rainwater harvesting simply means catching and holding rain where it falls and using it (http://www.rainwaterharvesting.org/index_files/FAQ.htm, accessed 16 August 2004). In different regions of the world, people have historically created catchment systems to harvest rainwater for human and animal consumption, cleaning purposes and agriculture. In regions where water is scarce, complex systems have been devised in order to save as much as possible of the precious liquid. The water management of a region provides a key to the understanding of the land use and related forces of production (Scarborough et al., 1995). The way the water was used can be observed through landscape modifications.

The Maya developed efficient methods of managing the available water. For example, in Palenque the presence of pools, an aqueduct and dams evidences an elaborate system of water management (French in Barnhart, 1999). In smaller Maya sites like La Milpa in Belize, the Maya created a micro-watershed with a reservoir-based water system to store runoff during the dry season. This system implies the alteration of the course of some streams to drain the ridge lands and at the same time slow water velocity, as well as the construction and/or use of drainages, dams, *aguadas*, *chultuno'ob* and reservoirs (Scarborough et al., 1995).

In Calakmul water is scarce in the dry season, but it can also be destructive in the rainy period. The *custodios* of Calakmul (the workers that take care of the archaeological site) informed me that during heavy rains the water level in the *corriental* rises more than two meters and the speed of the current is very destructive. Many dead animals and even fishes from distant places are dragged along. This phenomenon increases exponentially during hurricanes that periodically devastate the area. As Fray Diego de Landa affirms: "...because when it is time, vigorous and very strong rains fall and some times with many thunders and lightning" (de Landa, 1994:200, my translation), it is probable that the Maya of Calakmul protected themselves making sure that all this extra water could be diverted to reservoirs and flat areas.

The Maya made many arrangements for storing water. As Fray Diego de Landa wrote: "The Indians of the mountains, because of having very deep wells, they used to dig concavities in the rocks of their homes, and from there they pick up water from the rain" (de Landa, 1994:200, my translation). These concavities are called *chultuno'ob* (*chultun* in singular; see Figure 6.4). Twenty-six of them were identified in Calakmul (although some of them could be used for storing food).

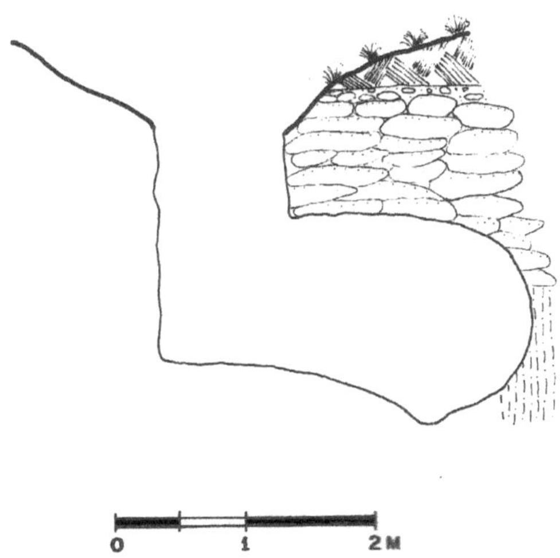

Figure 6.4. Cross section of a chultun (underground reservoir) with a simple chamber found in Calakmul (Zapata, 1985).

Considering the scarcity of water, how was it possible for ancient people to live with so little water? Possibly, they were able to cope with environmental circumstances with an integral system. If only the three biggest *aguadas* were used for the entire population, there certainly would be drastic water shortages. However the combined system, with many smaller *aguadas*, *chultuno'ob*, sinkholes and intermittent streams, potentially could have supported a larger population.

In order to divert and store water the people of Calakmul created a complex hydraulic system, a rain harvesting system. The primary advantage inherent in the physical setting of Calakmul is that all water collected from the eastern hills passes toward the north of the promontory following the water catchment that ends in El Laberinto *Bajo* (Figure 3.5).

As the water drains downslope, it is diverted through streams into the *corriental*, and then into reservoirs. In the same way, the rainwater falling in the main plaza of Calakmul is directed into reservoirs catchment areas. The presence of an adequate slope that directs the runoff and even the existence of drainages in palace complexes are just a couple of options at Calakmul for directing water to specific places in urban contexts. The stucco floors of some courtyards and palaces contribute to the re-direction of water.

At present, the quality of subterraneous water in the Calakmul Reserve is poor, due to an high content of calcium sulphate, which renders it unfit for human consumption, and probably this situation was similar in the past. Consequently, current inhabitants store rainwater in cisterns or *aguadas* (García Gil et al., 2002). Similarly, at sites in northern Guatemala the potable water is obtained from *aguadas*. Water supplies are not constant, and barely sustain a workforce of 150 for a two to three month field season. As water recedes the concentration of *Eschericia coli* (coliform bacteria) presents a significant health risk (Hansen et al., 2002). The same situation is encountered at Calakmul camp. The *aguada* is currently enough for an average of 100 people, but in dry years it is necessary to purchase water or pipe it in from outside sources.

Reservoirs were indispensable features of Maya hydrology. The construction of reservoirs at Tikal (Scarborough and Gallopin, 1991), El Mirador, (Matheny, 1986), La Milpa (Scarborough et al., 1995) or at Kinal (Scarborough et al., 1994) evidences the importance of rainfall dependent water management systems in the Maya Lowlands. The Maya inhabitants of the urban centers began constructing reservoirs and other water features toward the end of the Preclassic Period and also during the Classic. As Dunning and colleagues state (1999:656):

"Taking advantage of heavy rainfall for seven months of the year, the Maya could impound enough water in urban areas to survive the dry season. In the Preclassic, such reservoir systems typically occupied natural topographic depressions. However, as Classic Maya civilization took shape, reservoir construction shifted to the upper reaches of local watersheds to the rugged karst ridges on which urban centers were being built".

The study region, measuring 64.60 km^2, includes the upper part of El Laberinto *bajo*, the Calakmul promontory and a portion of the eastern hills (for a visual reference see Figure 6.3). In this area 98 reservoirs (natural and artificial) were counted. The approximate surface that they cover is 0.09 km^2. The biggest *aguada* has an area of 26 035.36 m^2; the other large ones measure 15 808.12 m^2 and 4453 m^2, respectively. If some of the reservoirs that were observed in stream meanders were also *aguadas*, then the surface area of potential water sources would increase a lot.

In contrast to what Pope and Dahlin (1989) affirm, through 1:20 000 ortho-photographs, lattices have been observed in the *Bajo* El Laberinto, as previously shown in Figure 3.9. In the same way, recent studies that employ remote sensing and field reconnaissance have established that at least some *akalche* portions were cultivated in ancient Maya sites (Beach et. al., 2003). It is important to note that many of the streams may have had a natural origin. As part of a superficial hydrological system typical of the Yucatan Peninsula, large streams receive water from smaller tributaries. It is obvious that the Maya could have taken advantage or modified natural features in order to more efficiently manage water.

During the 2004 season, a pit was dug in a channel that connects two large *aguadas* (Pit O; Figure 6.5). The channel, located in the north of the site center, measures about 250 meters of length. The profile of this channel had a black organic horizon followed by Layer I with dark structured soil, followed by another layer (II) composed of large rocks, each with an average size of thirty centimeters. These rocks were embedded in clayed gray sediment. Additionally, a dark gray clayed layer was the foundation of these rocks (Layer III). At approximately 70 cm of depth, the bedrock was reached. All of the layers had a strongly developed clay structure and were humid, a sign of poor drainage.

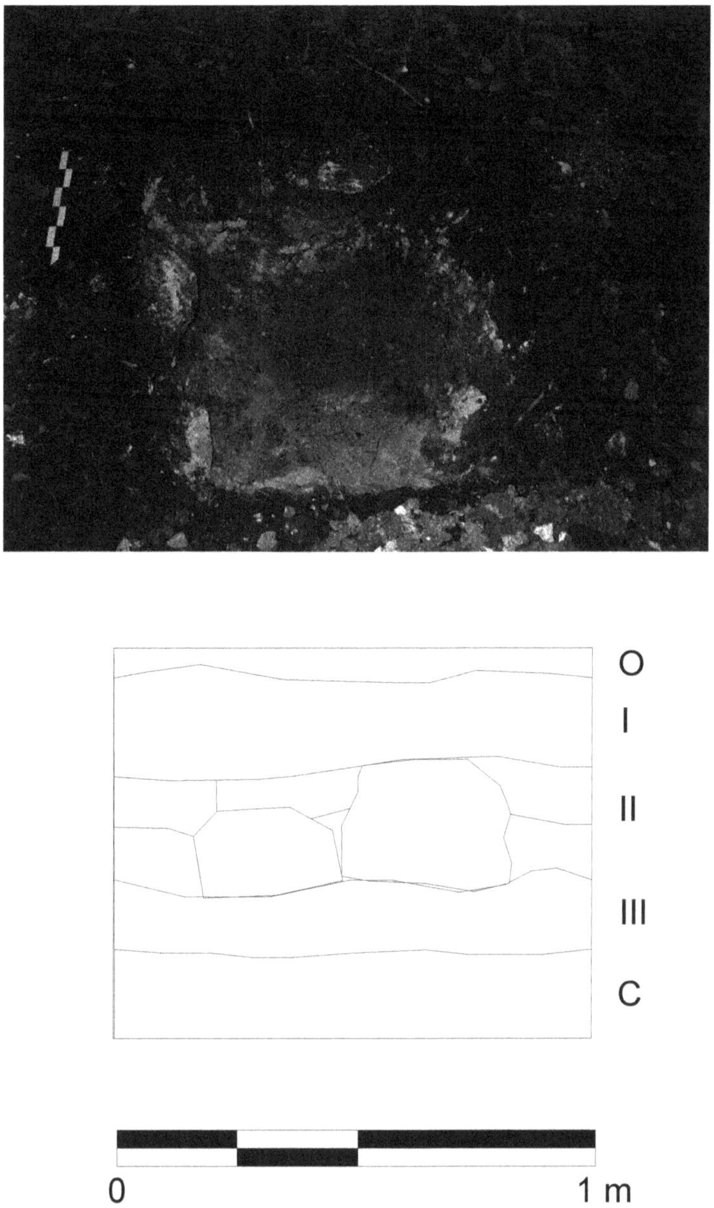

Figure 6.5. Channel image and profile. Pit O. O is organic horizon; I, II and III, the label levels. C is the slightly broken-up bedrock horizon.

It appears that these rocks, mixed with lithics and ceramics rubbish, formed a seal to impede water filtration and promote water channeling to specific places. The fact that the Maya realized these monumental works is all the more impressive considering the great length of some channels. The same type of seal was observed around *aguadas* (Pit P) with the same functional purpose: to retain water.

Canals are also common in many Maya sites. They have been observed in Belize (Pyburn, 2003; Siemens, 1982), along the upper Candelaria River, southern Quintana Roo (Pope and Dahlin, 1989), in Edzna, and along the Holmul River and its tributaries (Sever and Irwin, 2003).

City Occupation

- Ceramics

There are numerous publications that deal with the dynastic history of Calakmul (Carrasco, 2000; Carrasco et al., 1999b; Martin, 2000; Pincemin et al., 1998) and its interaction with neighboring sites (Martin and Grube, 2002). In contrast, little is known about the non-elite inhabitants of the city or kingdom: the commoners who worked the land and maintained the city.

Although this investigation was done following edaphological procedures, pottery and lithics were recovered. Only 10 pits had human vestiges. All ceramics and lithics from the 2003 and 2004 field seasons were excavated and carefully extracted according to natural horizons or artificial layers. The material was labeled, washed and marked by field workers and myself, and finally identified by archaeologist Alma Eugenia Martínez at the Ceramics Laboratory of the Calakmul Archaeological Project in Campeche.

The first study of the ceramics of Calakmul was done by archaeologist Rosario Domínguez (Domínguez Carrasco, 1994). This classification was made with the material from the excavation of structures I, II, III and VII. Subsequently, archaeologist Sylviane Boucher and colleagues analyzed the material form the excavations directed by archaeologist Ramon Carrasco in multiple buildings of the monumental area. They have reported materials form Middle Preclassic to Early Postclassic (e.g. Carrasco et al., 1996).

The classification of the ceramics of Calakmul follows the variety-type system. This system is based in attributes that are organized in different taxonomic categories as ware, group, type and variety, in combination with an analytic classification that includes modal atributtes and their distribution, in order to reflect traditions and stylistic horizons. The combination of these analyses give information related to chronology, ceramics distribution, interrelations and cultural processes (Boucher and Rojas, 1995). The present work considers the classification proposed by Boucher.

Ceramic material was found in pits B, C, H, I, J, K, L, M, N and O had (See Figure 5.1). Pits D and Q did not yield any ceramic remains. The ceramics dated to the Late Preclassic (17%), Early Classic (39%), Late Classic (24%), and Terminal Classic (20%) periods (Figure 6.6; expressed in quantities). No sherds were recovered from the Middle Preclassic or Postclassic periods. A total of 1037 specific, individual items were analyzed. Sixty sherds of that total could not be positively identified. None of these pits was excavated in evident plazas or buildings.

The elevated percentage of ceramics from the Early Classic coincides with the construction or modification of many monumental structures of the site. For example, the majority of burials and caches from structure IV were done during Early Classic (Carrasco and Boucher, 1994). During this same period, the construction of Sub-structure I-4 is done (Vázquez, 2006). However, this situation contrasts with the fact that few events were registered during Early Classic. It has been proposed that the lack of inscriptions for 109 years suggests the destruction of the capital of the Snake Kingdom or its movement to another place (Martin and Grube, 2002).

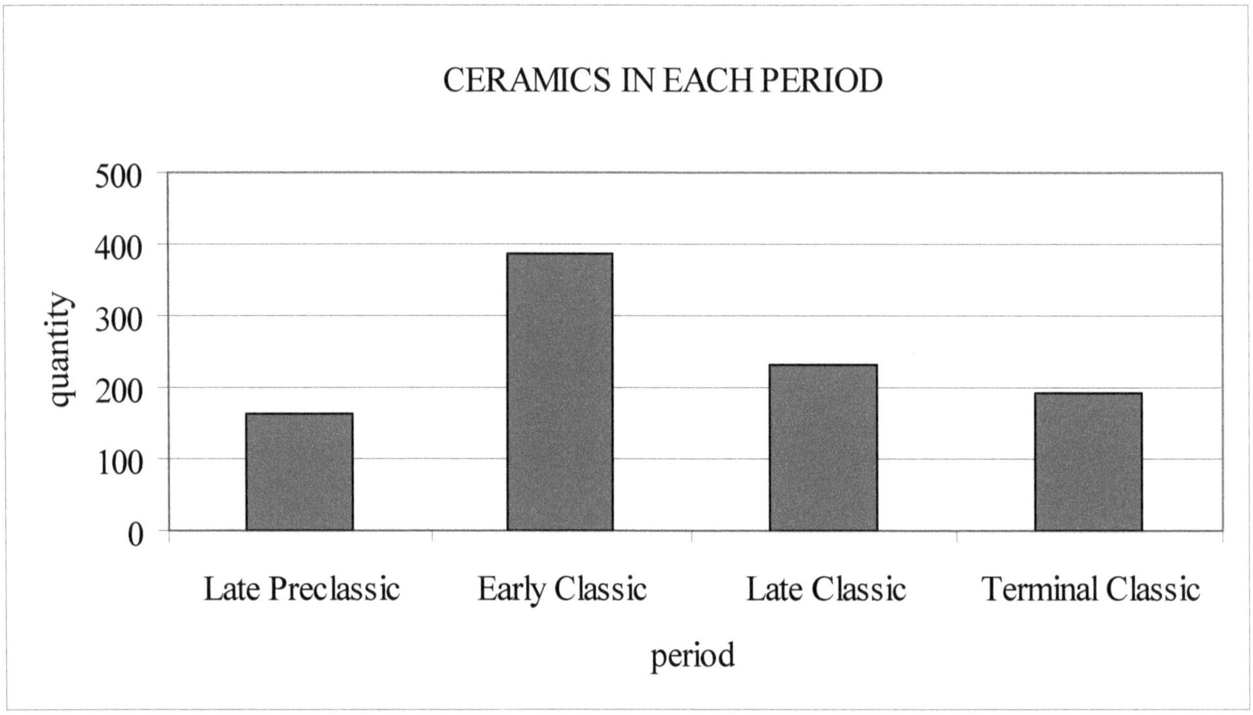

Figure 6.6. Total quantity of ceramics in all test pits during each Mesoamerican period.

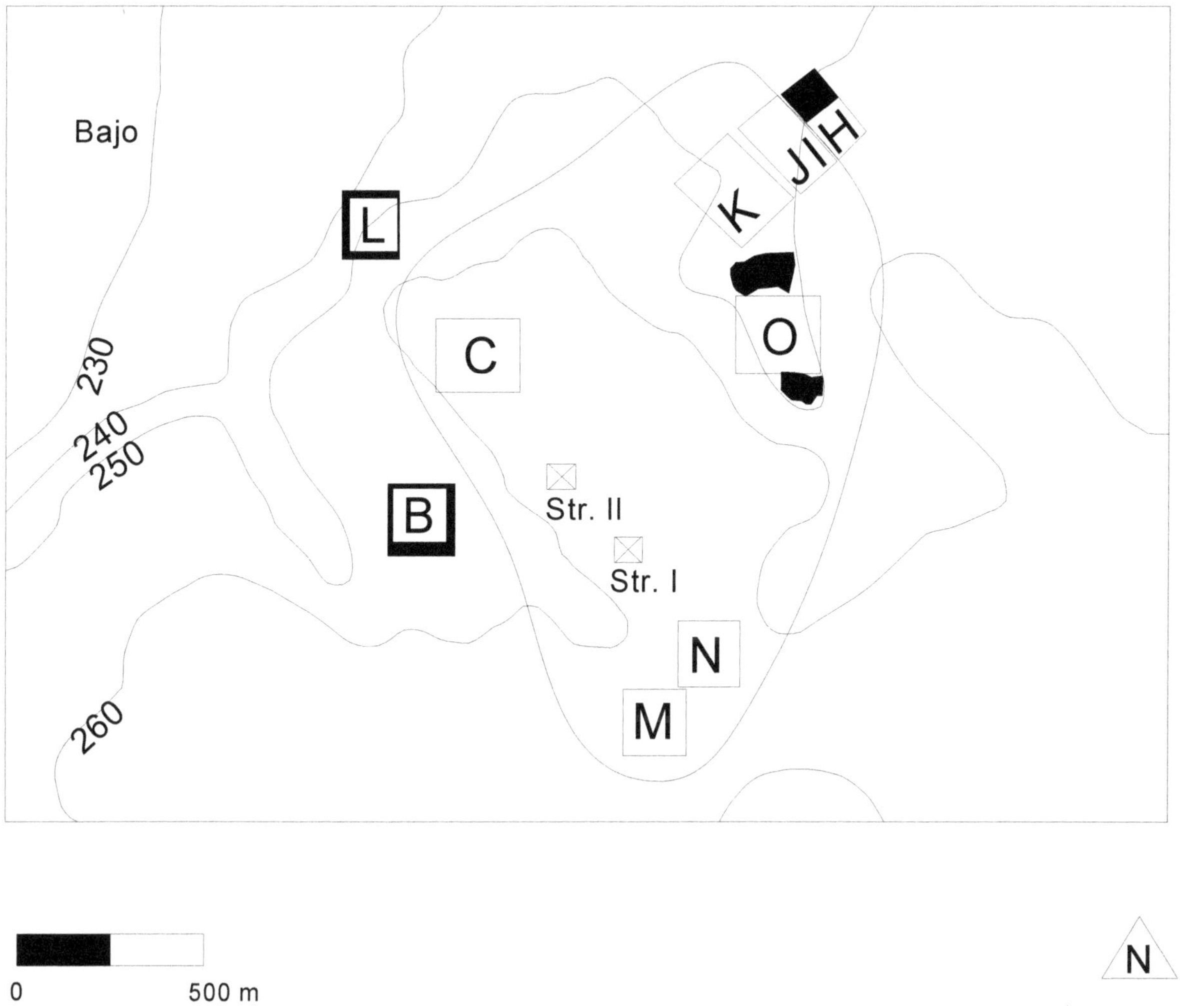

Figure 6.7. Ceramics temporality in each edaphological pit. Pits outside the curve had only Classic period ceramics. Str. means structure.

As can be appreciated in Figure 6.7, six pits contained ceramics that provenance from all mentioned temporal contexts (the ones inside the curve, from Late Preclassic to Terminal Classic). Pit O yielded material from the Late Preclassic, Early Classic and Late Classic; pit H from all phases of the Classic Period; pit B from the Early and Terminal Classic, and finally pit L from the Early and Late Classic periods.

Although this quantitative sample of pits is not enough to interpret ceramic distribution over the peripheral area of Calakmul, a potential pattern does emerge. The early material is closely concentrated around the civic-ceremonial center and near one of the *aguadas* (surrounded with a curve in Figure 6.7). Later material surrounds this area. No material appeared in the most distant pits located on the periphery. This observation suggests an early and incessant occupation very close to the area around the civic-ceremonial center of Calakmul. Subsequently, as population increased, more space or previously unoccupied land was required and consequently utilized during the Classic period. Much more detailed investigations should be realized in order to reliably confirm this interpretation.

The results of the ceramic analysis show that the main types of the Late Preclassic are Sierra rojo and Sapote estriado. Early Classic styles include Balanza negro, Águila naranja, Tacopate chorreado/café, Chiricote chorreado, Triunfo estriado and Quintal sin engobe. The most representative Late Classic wares are Infierno negro and Tinaja rojo nanzal. The main ceramic types of the Terminal Classic are Achote negro, Pizarra tanchacal and Tinaja rojo tinaja. The principal types and their percentage in the sample are shown in Table 6.6. All of the groups, types, and varieties encountered and distributed by horizons and layers can be consulted in Appendix 4.

The majority of the sherds belong to domestic contexts and with some exceptions, they are crude if compared with fine polychrome materials obtained from elite residential areas (Carrasco and Colón, 2005).

In addition, more ceramics were found next to areas of human inhabitation, such as in pits M, N, C and H. Multiple households could be observed near these pits. On the periphery, in areas of slopes and *bajos*, the density decreases (Figure 6.8). Although the quantity of ceramics is not representative and therefore unfit for interpretive generalizing, these data can be useful in terms of proposing preliminary hypotheses for subsequent investigations.

Period	Type	%
Late Preclassic	Sapote estriado Sapote	7.19
Late Preclassic	Sierra rojo Sierra	7.91
Early Classic	Aguila naranja Aguila	12.74
Early Classic	Balanza negro Balanza	3.49
Early Classic	Ciricote chorreado	1.95
Early Classic	Quintal sin engobe Quintal	5.14
Early Classic	Tacopate chorreado/café Tacopate	3.19
Early Classic	Triunfo Triunfo	1.54
Early Classic	Triunfo estriado Triunfo	8.73
Late Classic	Infierno negro Bolocantal	2.67
Late Classic	Tinaja rojo Nanzal	15.31
Terminal Classic	Achote negro	2.67
Terminal Classic	Pizarra tancachacal	1.85
Terminal Classic	Tinaja rojo Tinaja	10.59
All	Others	15.00

Table 6.6. Principal types of ceramic materials found in the test pits in Calakmul.

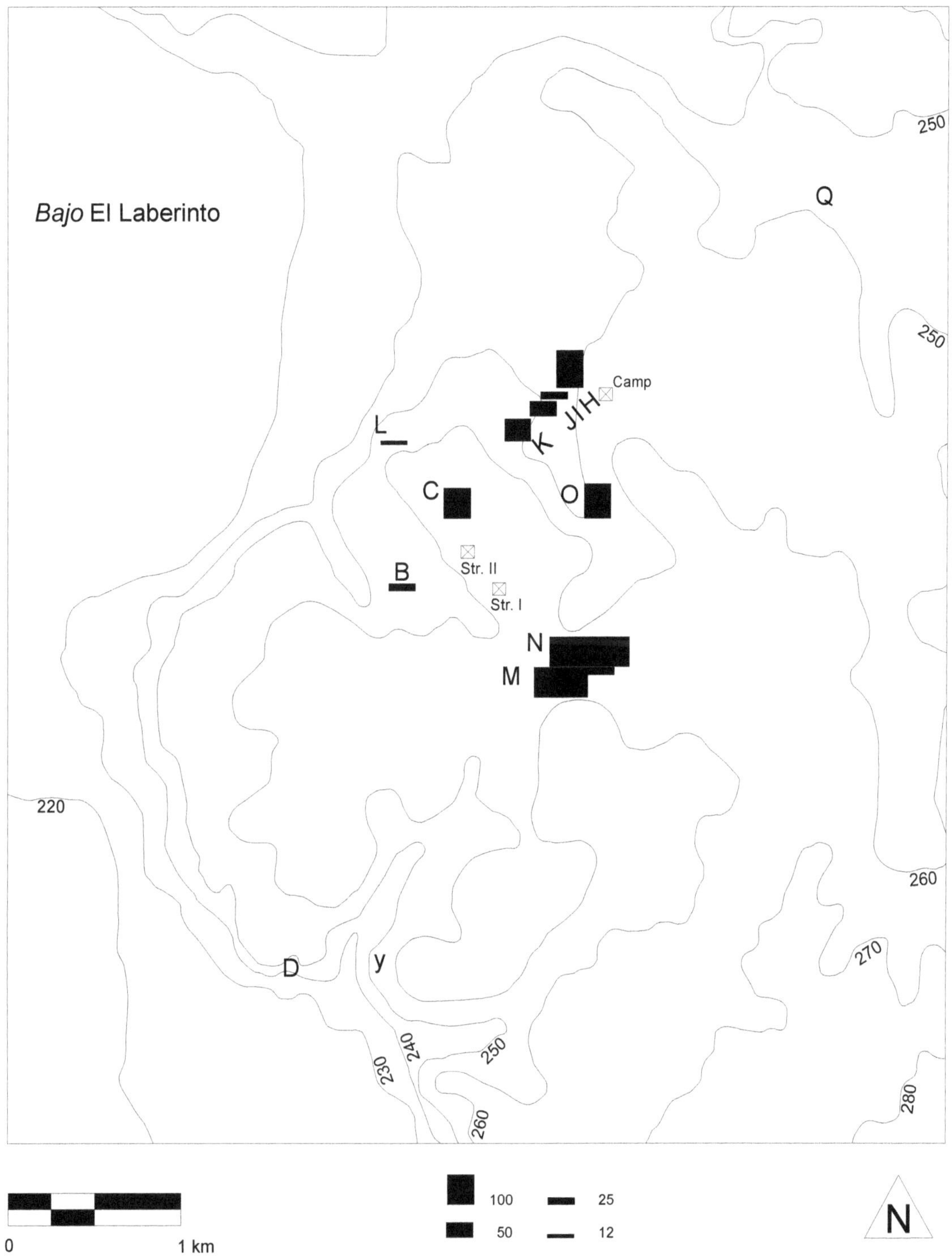

Figure 6 8. Ceramic density in each edaphological pit. The size of each square represents the quantity of sherds. The biggest square represents 100 sherds and the smallest represents 12. The squares are located next to the precise location of each pit, which is marked by a letter. Str. means structure.

- Lithics

Fifty tools, stones, and their fragments were found in eight pits in Calakmul. The great majority were flint fragments, some of them with cortex but with no sign of being worked. Lithics seemed to be discarded, previously broken, even burned and associated with refill (Figure 6.9). Lithic counts and identification can be appreciated in Appendix 4. The only recognizable tools were a limestone pick (Figure 6.1), a grinding stone (Figure 6.10), and a limestone tool adapted for a handle (Figure 6.11).

Figure 6.9. Flint fragments with and without cortex. The scale represents 3 cm.

Figure 6.10. Grinding stone. It has traces of a red pigment. The scale represents 3 cm.

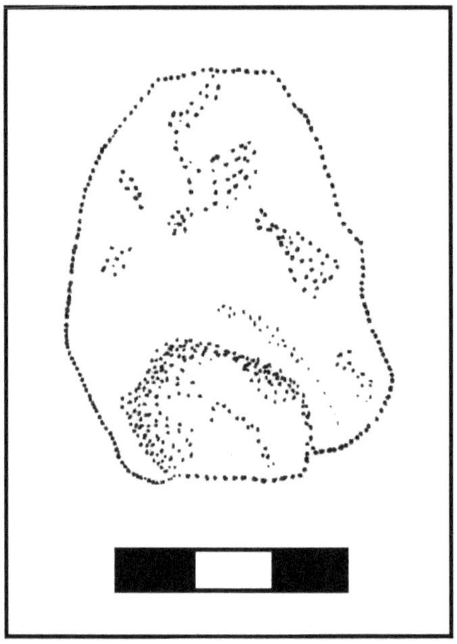

Figure 6.11. Limestone tool prepared for hafting in a wooden handle. The scale represents 3 cm.

- The Population of Calakmul

It is not clear yet which strategies were used to sustain the estimated population of around 20 000 people during the Classic Period (Fletcher and Gann, 1987), especially considering that the present inhabitants of Calakmul Reserve periphery barely subsist, cultivating thin soils and lacking permanent sources of fresh water. The yields of actual farmers have been calculated between 200-1500 kg/ha, being the mean of maize productivity of 500 kg/ha (Morales and Magaña, 2001). These low yields have lead to complete land abandonment of some of them after only three years (Escamilla et al., 2000).

Calculating the population of any archaeological site is a complex task. It is difficult to identify the function of the archaeological buildings, as well as their chronology or even the number of people who lived there. For these reasons demographic estimates are given based in suppositions and in ethnographic analogies.

Increasing evidence during the last two decades clearly points out to high population densities across the Maya lowlands (Culbert et al., 1990; Freter, 1994). Since Late Preclassic times the residents of the early settlements increased continually (Turner, 1990). During the Classic period the cities became large and almost every patch of terrain was occupied. This situation changed at the end of the Classic period, when the Maya lowlands were virtually abandoned.

In Calakmul kingdom, a similar demographic pattern has been elucidated (considering that the site's influence reached 5355 km^2; Turner, 1990:322; See Figure 6.12).

These first demographic estimates show that during the Middle Preclassic period about 79 000 people already lived there when presumably this site was a small settlement. During the Late Preclassic period, the population of the site got duplicated, and slightly augmented during the Early Classic to around 175 000. Impressively, during the Late Classic, population grew very fast until approximately 657 000 people. This growth did not last so much, since during Early and Late Postclassic periods it decreased to 120 000 and 22 000, respectively.

According to Turner, the population density before 300 B.C. was 15 persons/km^2 (Turner, 1990:322). In terms of Calakmul urban center, that equates to 424 persons in the promontory (within 3 kilometers radius), and a total of 754 including the rural area (within 5 km radius). The earliest population of Calakmul as a small settlement before 300 B.C. would be then 1178. Following these same calculations for the following periods, the total population during the Late Preclassic period was 2332, during the Late Classic 9668 and 330 during the Late Postclassic (Turner, 1990:322). Turner estimates for all the Mesoamerican periods are shown in Figure 6.13.

Culbert and Rice (1990) calculate the maximum population density in the Maya Lowlands during the Classic Period at 200 per km^2. Considering this estimate, the population of the Calakmul promontory area would be 5654 (within a radius of 3 km from the site center), and 15 708 in the 78.54 km^2 (within a radius of 5 km from the site center) during the Late Classic.

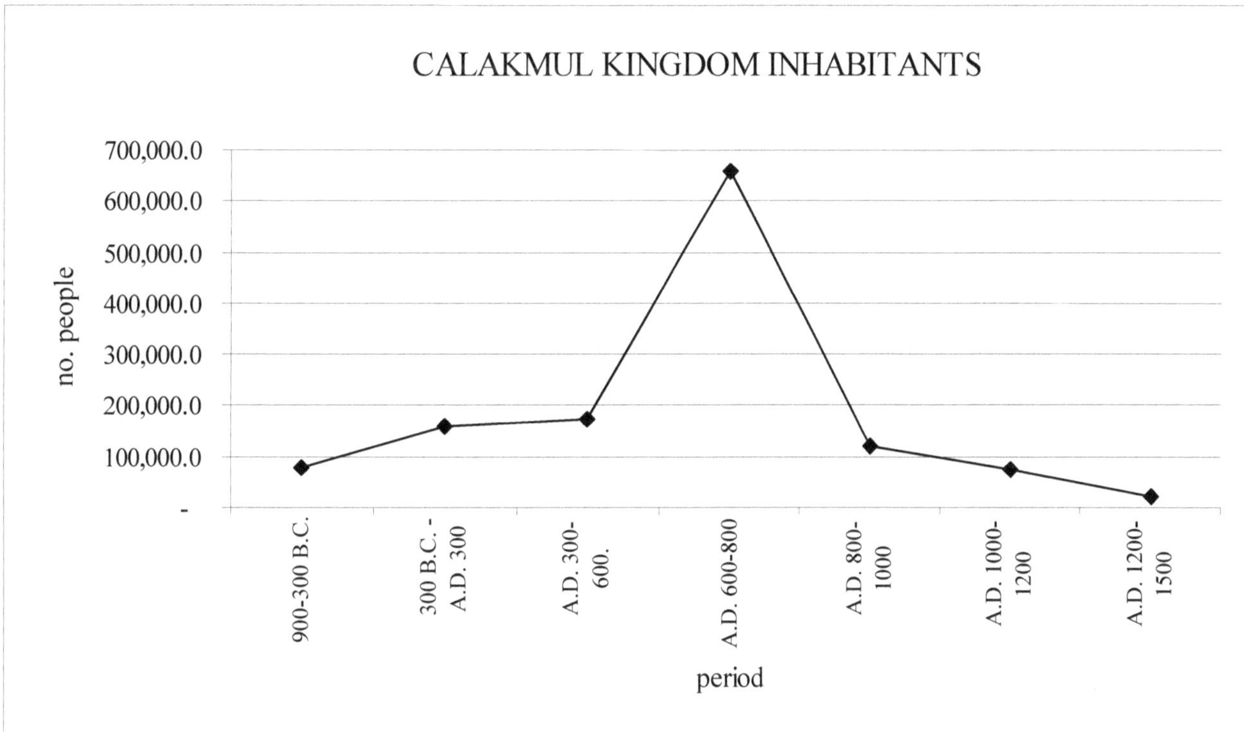

Figure 6.12. Inhabitant number of Calakmul kingdom (5355 km^2) since 900 B.C. to A.D. 1500 (based in Turner, 1990:322).

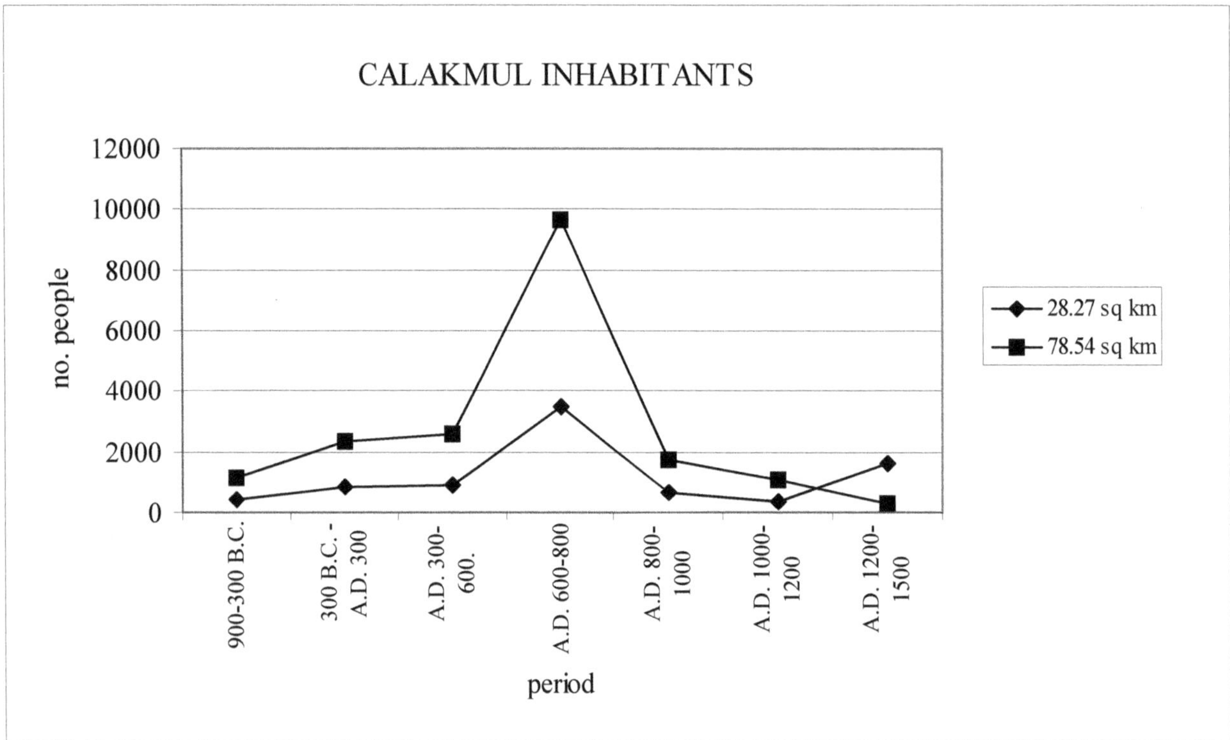

Figure 6.13. Population of Calakmul since 900 B.C. to A.D. 1500 in an area of 28.27 km^2 (within 3 km from site's center) and 78.54 km^2 (within 5 km from site's center), respectively (based in density/km^2 estimates; Turner, 1990:322).

Inhabitants/ 30 km^2	Formula	No. of buildings considered in calculations	References
15 475	Culbert	4863	Culbert et al., 1990; Fletcher and Gann, 1987
19 080	Folan	3407	Fletcher and Gann, 1987; Folan et al., 1983
15 854	Naroll	no	Fletcher and Gann, 1987
26 424	Naroll (Santley)	no	Fletcher and Gann, 1987; Santley, 1990:336

Table 6 .7. Inhabitants of Calakmul during the Late Classic Period. The formula and the number of buildings they used for calculating estimates are also shown (based in Fletcher and Gann, 1987; Folan et al., 2001a).

Other researchers have also calculated how much people lived in Calakmul (Fletcher and Gann, 1987; Folan et al., 2001a). After taking into account 6195 structures (presumably used for inhabitation and not including 105 altars) of the 30 km^2 map of Calakmul (May Hau et al., 1990), the following estimates have been reported for the Late Classic period (Table 6.7).

In sum, the population range of Calakmul during the Late Classic Period goes from 15 475 to 26 424 inhabitants (with a difference of 10 949, according to the formula they used). Folan and colleagues suggest a population of 50 000 inhabitants only in the city (Folan et al., 2001a:84). They have also reported other estimates for larger areas of influence. They affirm that the regional state of Calakmul (Braswell et al., 2004), including satellite sites (like Oxpemul, La Muñeca or Uxul), embraced a population of 200 000. Furthermore, they estimate that the rural population of Calakmul was of 1 500 000 inhabitants living on an area of 13 000 km^2.

In addition, carrying capacity studies provide another dimension to the population issue. Carrying capacity is established by the population density limits related to specific land uses (Turner et al., 2003). In Table 6.8 some estimates of Maya area carrying capacities are shown. It is evident that the environment of the northern Yucatan Peninsula supports less population in relation to areas and sites further south. Although these studies reflect current practices in the Maya region, they are useful in terms of suggesting potential explanations of the variability observed in estimates of population density.

It is relevant for this study to compare the demographic estimates between Calakmul and other ancient Maya sites. This information is very important for inferring the interaction between kingdoms, the development of cities and their degree of complexity. The population of Copan is relatively similar to the one of Calakmul (at least in the twenty thousands range; see Table 6.9). However, it is not easy to make comparisons with other sites since the estimates are reported with a varied space and time context.

As mentioned before, some demographic calculations for ancient sites have been obtained taking into account ethnographic studies. One excellent investigation of central Peten *milperos* in the vicinity of Lake Peten (Cowgill, 1961) proves useful in estimating ancient population. According to Cowgill, swidden farming was the most efficient agricultural system for the central Peten and most of the land was suitable for farming corn. She reports that swidden farming is comparably productive in the southern Maya Lowlands and that this system could support a population density of 100 to 200 people per square mile (38 to 77 people per km^2; Cowgill, 1961:283). Taking into account a similar density in Calakmul over a surface of 28.27 km^2 (as defined by the 3 kilometers radius, including the periphery of the site) the population would be 1074 to 2177. Over a surface of 78.54 km^2 (5 km radius, the inferred city limits), Calakmul would have a total population of 2984 to 6047, respectively.

Location	Production System[a]	Density people/km^2	Source
Northern Yucatán	Swidden *milpa*	19	Cook (Cook, 1972:31)
Northern Yucatán	Swidden *milpa*	23	Hester (Hester, 1954:129)
Copán Valley	Swidden *milpa*	25-32[b]	Wingard (Wingard, 1996)
Southern Petén	Swidden *milpa*	41[c]	Carter (Carter, 1969:142)
Greater Yucatán	Intensive swidden	57	Wagner (Wagner, 1968:185)
Central Petén	Swidden *milpa*	77[d]	Cowgill (Cowgill, 1961)

a. Swidden *milpa* = maize-based systems in which fallow exceeds cultivation (1:4 plus); Intensive swidden = maize based system in which length of fallow and cultivation are equal (1:1).
b. The lower figure for longer fallow (10 yr); the higher, for shorter fallow.
c. Figure or "available land" or cropped land, not total area of some bounded unit: noncropped land eliminated from estimate.
d. Figure assumes that virtually all land is high quality and suitable for maize cultivation. It is very high by world standards for swidden systems.

Table 6 8. Carrying capacity estimates for Maya Lowlands (quoted from Turner et al., 2003:369; source: Turner, 1983).

Site	Number of Inhabitants	Reference
Tikal (center plus outer)	13 275	Culbert and Rice 1990:116
Tikal periphery	45 720	Culbert and Rice 1990:116
Tikal total (120 km^2)	58 995	Culbert and Rice 1990:116
Copan	24 000	Freter, 1994
Piedras Negras epicenter	2600	Nelson, 2005:179
Piedras Negras polity	50 000	Nelson, 2005
Caracol	55 194 to 88 950	Murtha, 2002:143
Tayasal	21 951 to 32 272	Culbert and Rice, 1990
Chunchucmil	42 400 to 47 600	Dahlin et al., 2005
Yaxha	42 047	Culbert and Rice, 1990
Calakmul region (61 km^2)	20 000 to 30 000	Turner, 1990: 318,322
Calakmul	20 000	Fletcher and Gann, 1987

Table 6.9. Number of inhabitants of ancient Maya sites.

Similar numbers were obtained after calculating the number of inhabitants that swidden agriculture in the Calakmul study area could support (Chapter 5). It is proposed that depending on the number of crops per year, the maximum population that the promontory could support would be of 5637 to 11 961 inhabitants. As a consequence, there are two possible conclusions: Demographic estimations of Calakmul and other significant Maya sites are too high or that these large populations could only be supported with intensive agriculture practices.

It is more probable that these estimated population numbers may only be adequate for the population during the Middle Preclassic Period or the beginning of Late Preclassic, when Calakmul was a village in transition to an urban center. The estimations that were obtained following the land evaluation method are only for societies that do not practice intensive swidden agriculture. There is another example of a similar small population during the Late Classic in Maya area. Nelson affirms that epicentral Piedras Negras had a maximum population of 2600 inhabitants during the Chacalhaaz ceramic phase (A.D. 750 to A.D. 825; Nelson, 2005). It is very interesting to observe that this population could be maintained with a non-intensive medium fallow system.

The Agricultural Landscape of Calakmul

In sum, after analyzing the landscape of Calakmul it is possible to affirm that a considerable portion of the terrain was suitable for agriculture. The soils were capable for sustaining the *milpa* swidden system, root crops, cotton and other crops. The soils were deep enough as well as appropriate for good yields, and they could be improved with terrain modifications that imply irrigation and the use of natural fertilizers. Maya agriculturalists carried out practices that allowed the regeneration of soils and in conjunction with forest management, they were able to sustain large populations. They also took advantage of the wet period through a well-planned rain harvesting system. Water was kept in different types of reservoirs. The excessive liquid was conducted to them through drainages in the city, channels and modified streams in order to fill them and also to irrigate fields.

The landscape analysis, the results of the land evaluation, the information related with the settlement pattern (Folan et al., 2001a), the analysis of ceramics, former studies of prehispanic population as well as previous research in Calakmul allow me to suggest an hypothetical reconstruction of the ancient landscape of the site.

During the Late Preclassic period, the Maya of Calakmul settled gradually in the northwestern side of the promontory. The transition from perennial wetlands and shallow lakes to seasonal swamps favored their adaptation to the environment. They built the first civic-ceremonial monuments. They lived in stone platforms with houses made up of perishable materials, with gardens and small plots of *milpa* around them. They also planted in the margins of the swamps. Root crops and cotton were probably cultivated. They took a considerable advantage of the resources of the rainforest of the periphery. They kept water in reservoirs near the households. It is plausible that during the transition between the Late Preclassic and Classic a population of 5600 and 12 000 inhabited the site. A reconstruction of this landscape is seen in Figure 6.14. The drier conditions of the transition between Late Preclassic and Classic periods undoubtedly affected them, but those conditions did not lead into the abandonment of the city as it happened with many Mirador Basin sites.

During the Early and Late Classic periods the population augmented to about 20 000 individuals. Numerous structures, palaces and houses were built. They took advantage of all the available space in order to satisfy the primary needs of the growing population. The entire promontory was inhabited. The Maya of Calakmul developed a complex rain harvesting system that included the building and improvement of *aguadas*, the manipulation of streams to channel water into reservoirs in the city and in the *bajo* areas. These works were useful for keeping water from the rain, for inundating specific areas and for preventing flooding as well as great quantities of rain during the wet season. They increased the size of the *aguadas* with the objective of saving more water for the dry season. They cultivated *milpa* as well as other subsistence crops in the promontory, in transitional areas between the *bajos* and the promontory and in the *bajo* itself (drained fields). They had gardens surrounding the households and continued to take advantage of the resources of the surrounding rainforest. A reconstruction of the Classic Calakmul landscape is schematized in Figure 6.15.

The main necessities of the people of Calakmul were satisfied in a relatively nearby area from the site center: lithics in the *bajo*, clay for ceramics, limestone for construction, wood, fauna and flora form rainforest. Evenmore, if Calakmul was a major center of flint exploitation, resources derived from commerce with other settlements could be also valuable for sustaining people. But most likely their greater wealth was based in the knowledge of how to exploit their resources. This understanding implied a strict organization of the labor and a detailed observation of nature for generations.

The impressive architecture and elevated population density indicates a high degree of social organization, and implies that the commoners worked more land that was needed for their own subsistence. Subsequently, wetland agriculture was developed to feed the growing population of both elites and commoners. The arrival of a dry period during the end of the Classic impacted the environment considerably and witeventually led to the disappearance of the rainforest and finally of the people of Calakmul.

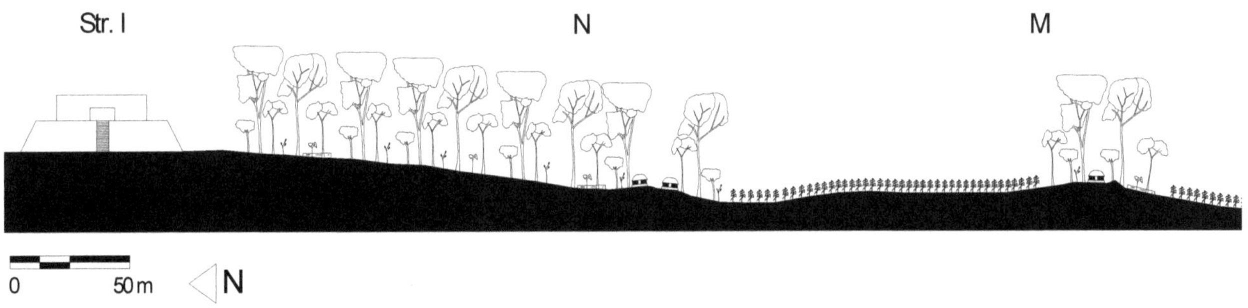

Figure 6.14. Hypothetical reconstruction of a Late Preclassic Calakmul landscape. The data were taken in the field. The capital letters above the drawing indicate the places where pits were excavated. The small buildings under trees represent households; the small plants, a milpa.

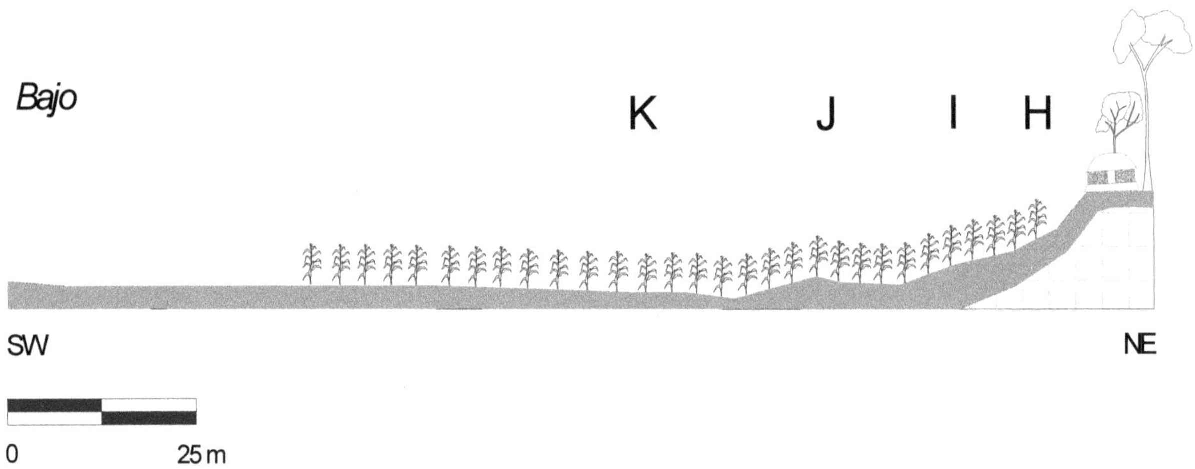

Figure 6.15. Hypothetical reconstruction of a Classic Calakmul landscape. The capital letters indicate the location of test pits. The data were taken in the field in a west-east transect behind the camp. The small building under trees represent a household; the small plants, a milpa

Work still to do at Calakmul

A more detailed study is required to observe if the areas surrounded by channels or streams are raised fields. The construction of raised platforms in wetland areas, even before the peak population of the Classic Period, was a way of taking advantage of the terrain in the Maya area. Experiments with raised fields in South America have shown that these features improve soil conditions (particularly nutrient content), provide localized drainage and the means for water management and organic recycling (Erickson, 1988).

In the same way a deeper study of the way the Maya of Calakmul caught water in the site center could be very interesting. There are enormous drainages between the building complexes and also within the palaces. A large-scale study of the water management of the urban center could refine the hypotheses about the way Maya rain harvested water.

It is also very important to investigate where and how non-elite people lived. The great majority of the excavations have been conducted in the monumental area, and little is know about the people who built and maintained this great city.

The Contribution of This Study

The Calakmul Biosphere Reserve is a crucial area since it regulates superficial and subterraneous hydrologic fluxes, replenishing the aquifers whose water reaches the most populated cities of the peninsula (García Gil et al., 2002). In addition, in conjunction with Guatemalan and other adjoining Mexican reserves, it is a significant refuge for wildlife. The forest itself also generates considerable quantities of oxygen.

In recent times, Calakmul rainforest has not been considered a suitable area for settlement due to the poor quality of subterraneous water and its scarcity during the dry season. Although a large number of hydraulic projects have been undertaken in an effort to provide a solution to the problem of water scarcity, at present the lack of permanent water sources still presents the major obstacle to social and economic development in the Calakmul region (García Gil et al., 2002). The present study suggests some alternatives, based on the methods employed by the ancient Maya to overcome the same difficulty.

The Calakmul Biosphere Reserve has long been a testing ground of sustainable agriculture. In 1991 an agroforestry system that integrates basic crops with fruit and forest exploitation was implemented, and it continues in operation today (Issac Márquez, 2004). Since the reserve was declared in 1989, many sustainable schemes have been suggested. For example, researchers have proposed that present day communal plots should be divided into units for optimal and efficient management (Acopa and Boege, 1998; Boege, 1993). Under this scheme, each plot would be subdivided for apiculture, permanentagriculture, forestry, cattle and intensive family orchards. Additionally, native rainforest products would be collected, such as *chicle* and allspice.

Based in this suggestion, it is proposed that a careful study of the areas in which the Maya chose to establish each of their agricultural practices could serve as a guide for setting up contemporary practices in the periphery of the Reserve. Until the present, wetlands have been considered wasted space because humans have rarely been able to take advantage of them. Uninformed, unchecked, or unsustainable development endangers this unique habitat, which shelters impressive orchids, scrubs, insects, birds and mammals, and that is now under great threat. For example, at present there is a project that is being developed to extract charcoal from *akalche* vegetation in communities of Quintana Roo (Joann Andrews, 2005, personal communication). Thousands of hectares will be burned in this process, turning *akalche* vegetation into a virtual desert. For centuries, the ancient Maya successfully adapted to these same habitats, surviving and thriving, at least until the Late Classic period. We can only hope that people and governments learn from the successes –and the mistakes– of the ancient Maya in dealing with the rainforest and *bajo* landscape of Calakmul

APPENDIX 1. - PROCEDURES FOR PREPARING AND ANALYZING SOILS

Collecting and Preparing Soil Sampling

(Modified from Foth, 1970:21).

1. A clean trowel, spoon or knife can be used to obtain the samples.
2. Collect four of five heaping teaspoons of soil.
3. Place the soil samples in the plastic sampling bag.
4. The soil should not be touched with the hands any more than is absolutely necessary.
5. After the sample is collected, it should be placed in a piece of clean paper or plastic. The soil should be allowed to dry in the air for several hours or overnight. Do not bake the soil to accelerate drying.
6. All foreign matter (leaves, roots, stones, etc.) should be removed.
7. All lumps should be crushed until the largest clumps are not larger than 2 mm diameter.

Color Protocol

The Munsell Soil Color Charts are the guidelines for classifying colors of soils, rocks and archaeological specimens, developed in cooperation with the U.S. Soil Conservation Service (Munsell Color Company Inc, 1954).

- Materials

1 white plain small container
1 set of soil Munsell Color Charts
1 dropper with water

- Procedure

Place a small quantity of soil in the white container. Determine the color comparing the soil with the most similar color of the Munsell Color Charts. Add 3 drops of water to the soil. Determine the color using the charts.

Soil pH Protocol

(based in Porta et al., 1994)

The pH scale is used as an indication of the concentration of hydrogen ions in the soil. The pH scale ranges from 1 to 14, with pH 1 being extremely acidic (1x10-1 moles of hydrogen ions per liter), pH 7 being neutral (1x10-7 moles of hydrogen ions per liter) and pH 14 being extremely basic (1x10-14 moles of hydrogen ions per liter). Soil pH indicates the chemical properties and fertility of the matrix and this is important because it constitutes an indicator of the plants that grow best in a specific place. In addition, pH varies according parent material, hydrology, biological activity and human influence.

- Materials

De-ionized water
Three 100 ml glass bottle per sample
Glass stirring rod
Spoon
Balance (accuracy to 0.1 g)
Calibrated pH meter

- Procedure

Weigh 10 g of soil in the balance and place it in the glass bottle. Add 25 ml of de-ionized water. Stir well for one minute with the glass rod. Rinse the rod with de-ionized water before stirring a different sample. Let the sample stand for an hour and measure it with the pH meter. Three tests from the same soil sample should be done. The reported pH result will be the average. The result will be expressed in terms of the pH scale values that go from 1 extremely acid to 14 extremely alkaline, being 7 neutral.

Soil Conductivity Protocol

(based in Aguilar, 1987; IRENAT, n. d.)

The conductivity is a measure of the total quantity of soluble salts of the soil.

- Materials

De-ionized water
Three 100 ml glass bottle per sample
Glass stirring rod
Spoon
Balance (accuracy to 0.1 gr)
Calibrated conductivity meter

- Procedure

Weigh 10 gr of soil in the balance and place it in the glass bottle. Add 50 ml of de-ionized water. Stir well for one minute with the glass rod. Rinse the rod with de-ionized water before stirring a different sample. Let the sample stand for an hour and measure it with the conductivity meter. Three tests from the same soil sample should be done. The reported conductivity result will be the average. The value of conductivity is expressed in miliSiemens por cm (mS/cm, or mmho each cm; 1 mmho/cm = 1 mS/cm)

Soil Texture Protocol

(LaMotte, n. d.)

The texture is the size of the particles of a determined soil. It is important to recognize the soil texture because it is a humidity and nutrient indicator. According to their size and after a separating the particles with the dispersion method, the texture of the particles is (LaMotte, n. d.):

The used text is designed to separate the soil in its three basic mineral fractions: sand, silt and clay (LaMotte, n. d.). The amount of time required for the soil particles of various sizes to settle in the soil separation tubes forms the basis for this test. From the amount of material collected in each tube it is possible to determine the approximate percentage of each fraction as represented in the original soil sample.

Particle	Size in mm
very coarse sand	2.0-1.0
coarse sand	1.0-0.5
medium-sized sand	0.5-0.25
fine sand	0.25-0.1
very fine sand	0.1-0.05
silt	0.05-0.002
clay	<0.002

- Materials
1 ml of Soil Flocculating Reagent
1 ml of Texture Dispersing Reagent
1 soil texture stand
3 test tubes, conic, specially designed for this test, 50 ml with cap, marked A, B and C
2 pipettes, 1 ml, plastic, with cap

- Procedure
Place the three soil separation tubes in the rack. Add the soil sample to soil separation tube A. Gently tap the bottom of the tube on a firm surface to pack the soil and eliminate air spaces. Use the pipette to add 1 ml of texture dispersing reagent to the sample in soil separation tube A. Dilute to line 45 with tap water. Cap and gently shake for two minutes, making sure all the soil sample is thoroughly mixed with water. The sample is now ready for separation. The separation is accomplished by allowing a predetermined time for each fraction to settle out of the solution. Be sure that you continue to gently shake the separation tube up to the time of the first separation. Place soil separation tube A in the rack. Allow to stand undisturbed for exactly 30 seconds. Carefully pour off all the solution into Soil Separation Tube B. Return Tube A to the rack. Allow tube B to stand undisturbed for 30 minutes. Carefully pour off the solution from soil separation tube B into Soil Separation Tube C. Return Tube B to the rack. Add 1 ml of soil flocculation reagent to soil separation tube C. Cap and gently shake for one minute. Place the soil separation tube C in the rack and allow to stand until all the clay in suspension settles. This may require up to 24 hours (LaMotte n. d.).

Due to the colloidal nature of clay in solution and its tendency to swell and form a gel, the portion of clay remaining in tube C is not used to determine the clay fraction present in the soil. The clay fraction is calculated by adding the sand and silt fractions, and subtracting this total from the initial volume of soil used for separation. So, read soil separation tube A at top of soil level. To calculate percentage sand in the soil, divide reading by 15. Multiply by 100. Record as % sand. Then read soil separation tube B at top of soil level. To calculate percentage silt in the soil, divide reading by 15. Multiply by 100. Record as % silt. Calculate volume of clay as shown above. To calculate percent clay n the soil, divide the value by 15. Multiply by 100. Record as % clay (LaMotte, n. d.).

Sandy soil is described as soil material that contains 85% or more sand. The percentage of silt plus 1.5 times the percentage of clay shall not exceed 15. Silt soil is described as soil material that contains 80% or more silt and less than 12% clay. Clay soil is described as soil material that contains 40% or more clay, les than 45% sand and les than 40% silt (LaMotte, n. d.).

Organic Matter Protocol
(Nuñez, 1977)

- Materials
Furnace with temperature range between 50 and 450 ºC
Scale
1 crucible
Desiccation container

- Procedure
Put each crucible in the furnace with 400 ºC for two hours. Let it get cool in the desiccation container. Place 5 grams in the furnace with 110 ºC for two hours. Put it in the desiccation container and weigh the sample. Place the crucible with the sample in the furnace with 400 ºC for eight hours. Weigh the crucible when it is cool. Calculate the loss of weigh percentage during ignition according the next formula:
Loss organic material by ignition = Dry weight – Ignition weight / Dry weight – Crucible weight

Hanna Mehlich Phosphorus Extraction Procedure and Phosphorus Test for Soil

This method Mehlich-2 for extracting phosphorus is similar to Mehlich-3 but not so commonly used in United States of America (Heckman, n.d.).

- Materials
100 ml (1 bottle) of Mehlich-2 extraction solution
Demineralized water
1-2 mm soil sieve
1 plastic test tube (14 ml) with screw cap
1 small funnel
Filter paper discs 80 mm diameter
1 tube stand
1 calibrated vessel (20 ml) with cap
1 sample cup (1 g)
1 plastic pipette (1 ml)
1 brush
1 spoon
HI 93713-0 reagent (1 packet per sample)
1 checker disk with the 38073 disk
2 glass vials with caps
1 syringe (1 ml) with tip

- Procedure
Use the spoon to fill the 1 g sample cup with the sieved soil sample and level the sample in the cup by discarding

the excess soil with the spoon handle. Using the plastic pipette, add to the plastic test tube 1 ml of extraction solution. Remove the cap and fill the demineralizer bottle with tap water. Replace the cap and shake gently for at least two minutes. The demineralized water is now ready. Flip open the top o the demineralizer bottle cap. By gently squeezing the bottle, add demineralized water to the test tube up to the 10 ml mark. Replace the screw cap and shake the test tube to mix. Remove the cap and add the soil sample from the sample cup to the tube. Replace the cap and mix by shaking the tube up and down for five minutes. Place the cap on the calibrated vessel and insert the funnel in the cap hole. Fold a filter paper disc twice to form a cone. Place the folded filter in the funnel and filter the sample. The extracted sample in the beaker is now ready for analysis.

Using the syringe, add to each glass vial 0.75 ml of sample. Add demineralized water to each glass vial up to the 7.5 mark, replace caps and swirl to mix. Remove caps and insert one of the vials into the left hand opening of the checker disc. This is the blank. Add to the other glass vial 1 packet of HI 93713-0 reagent. Replace the cap and shake vigorously for 20 seconds. This is the reacted sample. Wait for three minutes. Remove the cap and insert the reacted sample into the right hand opening of the checker disc. Hold the checker disc so that a light source illuminates the samples from the back of the windows. Keep the checker disc at a distance of 30-40 cm to match the color. Rotate the disc while looking at the color test windows and stop when you find the color match. Read the value in the result window, multiply by 3.3 and record it in mg/l (=ppm) of Phosphorus. For best results perform the reading three times and take the average value.

APPENDIX 2. FORMS FOR ENVIRONMENT AND SOIL PROFILE DESCRIPTION

Description of the Environment of the Soil Profile
(Modified from USDA SSS, 1962)

Soil Type: Date:
Estimated area: Profile: ID:
Classification: Form No.
Location: Observer:

Vegetation:
Soil Use:
Parent Material:
Landform:
Relief:
Stoniness:
Rocky Outcrop:
Roots:
Elevation:
Slope and Direction:
Erosion
Sedimentation
Permeability
Human Activities:
Atmospheric Time:
Climate:
Water Sources:
Humidity:

Photographs:

Observations:

Horizon Description
(Modified from USDA SSS, 1962)

Profile ID:
Form No.:
Bag No.:
Date:
Observer:
Location:

Horizon ID: Horizon __ Horizon _
Depth:
Color:
Structure:
Stoniness:
Roots:
Drainage:
Vegetation:
Fauna:
Photographs:
Observations:
Drawings:

APPENDIX 3. PHYSICAL AND CHEMICAL ANALYSES OF CALAKMUL SOILS

Physical Analyses of Calakmul Soil Profiles						2003 AND 2004
ID BAG	PIT	HORIZON	THICKNESS	DRY COLOR	DRY COLOR NAME	WET COLOR
1	A	O	0-3	10YR2/1	BLACK	10YR2/1
2	A	A	3-33	10YR3/2	VERY DARK GRAYISH BROWN	10YR4/1
3	A	Bk	33-58	10YR5/2	GRAYISH BROWN	10YR6/1
4	A	Bk/C	58-?	10YR6/2	LIGHT BROWNISH GRAY	10YR7/2
5	B	A	9-38	10YR3/2	VERY DARK GRAYISH BROWN	10YR2/2
6	C	O	0-3	7.5YR4/2	VERY DARK GRAYISH BROWN	10YR3/2
7	C	A	3-30	7.5YR4/2	VERY DARK GRAYISH BROWN	10YR3/2
8	C	Bk	30-58	10YR5/2	GRAYISH BROWN	10YR4/2
9	C	C	58-?	7.5YR6/1	GREY	7.5YR5/1
10	D	A	3-13	10YR3/2	VERY DARK GRAYISH BROWN	10YR2/1
11	H	O	0-10	10YR3/2	VERY DARK GRAYISH BROWN	10YR2/1
12	H	A	10-30	10YR4/1	DARK GREY	10YR3/2
13	H	Bk	30-53	10YR4/2	DARK GRAYISH BROWN	10YR3/1
14	I	O	0-10	7.5YR2.5/1	BLACK	7.5YR2.5/1
15	I	A	10-40	10YR2/1	BLACK	7.5YR2.5/1
16	I	Bk	40-60	10YR3/2	VERY DARK GRAYISH BROWN	10YR4/1
17	I	Bk/C	60-83	10YR5/1	GREY	10YR4/1
18	J	O	0-7	10YR2/1	BLACK	7.5YR2.5/1
19	J	A	7-30	10YR3/1	VERY DARK GREY	10YR3/1
20	J	Bk	30-62	10YR4/1	DARK GREY	10YR4/1
21	K	A	1-11	7.5YR3/1	VERY DARK GREY	7.5YR2.5/1
22	K	A/Bg	11-26	10YR5/1	GREY	10YR5/1
23	K	Bg	26-66	10YR6/2	LIGHT BROWNISH GRAY	10YR7/1
24	K	Bg/C	66-?	2.5Y5/2	GRAYISH BROWN	2.5Y6/2
25	L	A	6-16	10YR3/2	VERY DARK GRAYISH BROWN	10YR3/2
26	L	Bk	16-30	10YR5/2	GRAYISH BROWN	10YR4/2
27	L	C	30-?	10YR8/2	VERY PALE BROWN	10YR7/3
28	M	A	10-18	10YR3/1*	VERY DARK GREY*	10YR3/2*
29	M	Bk	18-21	7.5YR6/1	GREY	10YR4/1
30	N	A	9-27	7.5YR3/1	VERY DARK GREY	10YR3/2
31	N	Bk	27-86	7.5YR5/1*	GREY*	10YR4/1*
32	O	O	0-6	7.5YR2.5/1	BLACK	10YR2/1
33	O	I	6-26	10YR5/1	GREY	10YR4/1
34	O	II	26-40	10YR6/1	GREY	10YR5/1
35	O	III	40-?	10YR5/1	GREY	10YR6/1
36	P	I	5-15	7.5YR2.5/1	BLACK	7.5YR3/1
37	P	II	15-24	7.5YR5/1	GREY	10YR4/1
38	P	IV	24-?	10YR4/1	DARK GREY	10YR3/1
39	Q	O	0-4	10YR4/1	DARK GREY	10YR3/1
40	Q	A	4-12	10YR4/1	DARK GREY	10YR3/2
41	Q	Bg	12-?	10YR4/1	DARK GREY	10YR4/1
42	Str. II	RELL	—	10YR7/1*	LIGHT GREY	10YR4/1*
43	Str. II	RELL	—	7.5YR8/1	WHITE	10YR8/2
44	Str. II	RELL	—	7.5YR8/1	WHITE	10YR8/2
45	Str. II	RELL	—	10YR5/1	GREY	10YR4/1
46	Str. II	NO	—	7.5YR8/1	WHITE	10YR7/2
47	Wit-ness	NO	—	GLEYN4/	DARK GREY	10YR2/1

ID BAG	WET COLOR NAME	STRUCTURE DEVELOPMENT	STRUCTURE CLASS	STRUCTURE TYPE	STRUCTURE KIND
1	BLACK	WEAK	POLYHEDRAL	BLOCKY	SUBANGULAR
2	DARK GREY	MODERATE	POLYHEDRAL	BLOCKY	SUBANGULAR
3	GREY	MODERATE	POLYHEDRAL	BLOCKY	SUBANGULAR
4	LIGHT GREY	MODERATE	POLYHEDRAL	BLOCKY	SUBANGULAR
5	VERY DARK BROWN	WEAK	POLYHEDRAL	BLOCKY	SUBANGULAR
6	VERY DARK GRAYISH BROWN	WEAK	POLYHEDRAL	BLOCKY	SUBANGULAR
7	GRAYISH BROWN	WEAK	POLYHEDRAL	BLOCKY	SUBANGULAR
8	DARK GRAYISH BROWN	WEAK	POLYHEDRAL	BLOCKY	SUBANGULAR
9	GREY	WEAK	POLYHEDRAL	BLOCKY	SUBANGULAR
10	BLACK	WEAK	POLYHEDRAL	BLOCKY	SUBANGULAR
11	BLACK	WEAK	POLYHEDRAL	BLOCKY	SUBANGULAR
12	GRAYISH BROWN	WEAK	POLYHEDRAL	BLOCKY	SUBANGULAR
13	VERY DARK GREY	WITHOUT	STRUCTURELESS	AMORPHOUS	N/A
14	BLACK	WEAK	POLYHEDRAL	BLOCKY	SUBANGULAR
15	BLACK	WEAK	POLYHEDRAL	BLOCKY	SUBANGULAR
16	DARK GREY	WEAK	POLYHEDRAL	BLOCKY	SUBANGULAR
17	DARK GREY	WEAK	POLYHEDRAL	BLOCKY	SUBANGULAR
18	BLACK	WEAK	POLYHEDRAL	BLOCKY	SUBANGULAR
19	VERY DARK GREY	MODERATE	POLYHEDRAL	BLOCKY	SUBANGULAR
20	DARK GREY	MODERATE	POLYHEDRAL	BLOCKY	SUBANGULAR
21	BLACK	MODERATE	POLYHEDRAL	BLOCKY	SUBANGULAR
22	GREY	MODERATE	POLYHEDRAL	BLOCKY	SUBANGULAR
23	LIGHT GREY	STRONGLY DEVELOPED	POLYHEDRAL	BLOCKY	SUBANGULAR
24	LIGHT BROWNISH GRAY	STRONGLY DEVELOPED	POLYHEDRAL	BLOCKY	SUBANGULAR
25	VERY DARK GRAYISH BROWN	MODERATE	POLYHEDRAL	BLOCKY	SUBANGULAR
26	DARK GRAYISH BROWN	MODERATE	POLYHEDRAL	BLOCKY	SUBANGULAR
27	VERY PALE BROWN	WITHOUT	STRUCTURELESS	SINGLE GRAIN	N/A
28	VERY DARK GRAYISH BROWN*	MODERATE	POLYHEDRAL	BLOCKY	SUBANGULAR
29	DARK GREY	MODERATE	POLYHEDRAL	BLOCKY	SUBANGULAR
30	VERY DARK GRAYISH BROWN	WEAK	POLYHEDRAL	BLOCKY	SUBANGULAR
31	DARK GREY*	WITHOUT	STRUCTURELESS	AMORPHOUS	N/A
32	BLACK	MODERATE	POLYHEDRAL	BLOCKY	SUBANGULAR
33	DARK GREY	STRONGLY DEVELOPED	POLYHEDRAL	BLOCKY	SUBANGULAR
34	GREY	STRONGLY DEVELOPED	POLYHEDRAL	BLOCKY	SUBANGULAR
35	GREY	STRONGLY DEVELOPED	POLYHEDRAL	BLOCKY	SUBANGULAR
36	VERY DARK GREY	MODERATE	POLYHEDRAL	BLOCKY	SUBANGULAR
37	DARK GREY	MODERATE	POLYHEDRAL	BLOCKY	SUBANGULAR
38	VERY DARK GREY	MODERATE	POLYHEDRAL	BLOCKY	SUBANGULAR
39	VERY DARK GREY	MODERATE	POLYHEDRAL	BLOCKY	SUBANGULAR
40	VERY DARK GRAYISH BROWN	MODERATE	POLYHEDRAL	BLOCKY	SUBANGULAR
41	DARK GREY	STRONGLY DEVELOPED	POLYHEDRAL	BLOCKY	SUBANGULAR
42	DARK GREY	WITHOUT	STRUCTURELESS	AMORPHOUS	N/A
43	VERY PALE BROWN	WITHOUT	STRUCTURELESS	AMORPHOUS	N/A
44	VERY PALE BROWN	WITHOUT	STRUCTURELESS	AMORPHOUS	N/A
45	DARK GREY	WITHOUT	STRUCTURELESS	AMORPHOUS	N/A
46	LIGHT GREY	STRONGLY DEVELOPED	POLYHEDRAL	BLOCKY	SUBANGULAR
47	BLACK	WITHOUT	WITHOUT	N/A	N/A

ID BAG	% SAND	%SILT	%CLAY	TEXTURAL CLASS
1	6.67	53.33	40.00	SILTY CLAY
2	6.67	5.00	88.33	CLAY
3	30.00	13.33	56.67	CLAY
4	33.33	1.67	65.00	CLAY
5	8.33	66.67	25.00	SILTY CLAY LOAM
6	10.00	40.00	50.00	SILTY CLAY
7	21.67	30.00	48.33	CLAY
8	33.33	10.00	56.67	CLAY
9	26.67	16.67	56.67	CLAY
10	60.00	40.00	0.00	SANDY LOAM
11	6.67	56.67	36.67	SILTY CLAY LOAM
12	13.33	3.33	83.33	CLAY
13	33.33	26.67	40.00	CLAY LOAM
14	26.67	40.00	33.33	CLAY LOAM
15	23.33	23.33	53.33	CLAY
16	13.33	26.67	60.00	CLAY
17	23.33	46.67	30.00	CLAY LOAM
18	3.33	43.33	53.33	SILTY CLAY
19	36.67	30.00	33.33	CLAY LOAM
20	16.67	3.33	80.00	CLAY
21	1.67	3.33	95.00	CLAY
22	1.67	20.00	78.33	CLAY
23	13.33	10.00	76.67	CLAY
24	3.33	53.33	43.33	SILTY CLAY
25	1.67	33.33	65.00	CLAY
26	3.33	53.33	43.33	SILTY CLAY
27	10.00	33.33	56.67	CLAY
28	20.00	53.33	26.67	SILTY CLAY LOAM
29	23.33	20.00	56.67	CLAY
30	20.00	66.67	13.33	SILT LOAM
31	26.67	46.67	26.67	CLAY LOAM
32	3.33	3.33	93.33	CLAY
33	6.67	30.00	63.33	CLAY
34	3.33	5.00	91.67	CLAY
35	3.33	20.00	76.67	CLAY
36	33.33	33.33	33.33	CLAY LOAM
37	73.33	3.33	23.33	SANDY CLAY LOAM
38	6.67	6.67	86.67	CLAY
39	13.33	6.67	80.00	CLAY
40	3.33	3.33	93.33	CLAY
41	3.33	10.00	86.67	CLAY
42	36.67	16.67	46.67	CLAY
43	63.33	13.33	23.33	SANDY CLAY LOAM
44	66.67	13.33	20.00	SANDY CLAY LOAM
45	20.00	5.00	75.00	CLAY
46	86.67	5.00	8.33	LOAMY SAND
47	3.33	1.67	95.00	CLAY

ID BAG	APPARENT TEXTURE	CLAY CONCRETIONS	ROOTS	STONE %	VEGETATION	FAUNA	APPARENT DENSITY
1	CLAYISH	WITH	REGULAR	5	LEAVES, TRUNKS	BEETLE, SNAIL	1.01
2	CLAYISH	WITH	REGULAR	3	NO	SNAIL	1.01
3	CLAYISH	WITH	SCARCE	5	NO	NO	1.02
4	CLAYISH	WITH	NO	10	NO	NO	1.01
5	CLAYISH	MICRO-LIMESTONE	ABUNDANT	25	TRUNKS	SNAIL	1.00
6	SILT-SANDY	WITH	REGULAR	25	NO	SNAIL	1.01
7	SILT-SANDY	WITH	REGULAR	25	SEEDS	SNAIL	1.01
8	CLAY-SANDY	WITH	REGULAR	40	NO	NO	1.00
9	CLAY-SANDY	WITH	SCARCE	50	LEAVES, TRUNKS, SEEDS	SNAIL	1.01
10	SANDY	WITH	ABUNDANT	30	TRUNKS, SEEDS	SNAIL	1.01
11	SILTY	WITH	ABUNDANT	30	TRUNKS	SNAIL	1.00
12	SILTY	MICRO-LIMESTONE	ABUNDANT	40	NO	SNAIL	1.00
13	SILTY	MICRO-LIMESTONE	SCARCE	30	TRUNKS	NO	1.00
14	SILTY	MICRO-LIMESTONE	VERY ABUNDANT	30	TRUNKS	SNAIL	1.00
15	CLAYISH	WITH	ABUNDANT	25	NO	SNAIL	1.01
16	CLAY-SANDY	WITH	REGULAR	30	NO	SNAIL	1.01
17	CLAY-SANDY	WITH	REGULAR	50	NO	NO	1.00
18	CLAYISH	WITH	VERY ABUNDANT	3	TRUNKS	SNAIL	1.01
19	CLAYISH	WITH	ABUNDANT	5	NO	NO	1.01
20	CLAYISH	WITH	REGULAR	5	LEAVES	NO	1.02
21	CLAYISH	WITH	ABUNDANT	3	LEAVES, TRUNKS	NO	1.01
22	CLAYISH	WITH	REGULAR	3	NO	NO	1.01
23	CLAYISH	WITH	SCARCE	15	NO	NO	1.01
24	CLAYISH	WITH	NO	3	NO	NO	1.01
25	SILT-SANDY	WITH	VERY ABUNDANT	3	TRUNKS	SNAIL	1.01
26	CLAYISH	WITH	ABUNDANT	10	NO	SNAIL	1.01
27	SILTY	WITH	SCARCE	35	NO	NO	1.02
28	SILT-SANDY	WITH	REGULAR	40	NO	SNAIL	1.00
29	SILT-SANDY	WITH	NO	50	NO	SNAIL	1.01
30	CLAYISH	WITH	ABUNDANT	25	SEEDS, TRUNKS	SNAIL	1.01
31	SILT-SANDY	WITH	SCARCE	50	NO	NO	1.01
32	CLAYISH	WITH	ABUNDANT	25	LEAVES, TRUNKS	SNAIL	1.00
33	CLAYISH	WITH	REGULAR	10	NO	SNAIL	1.00
34	CLAYISH	WITHOUT	ABUNDANT	15	NO	NO	1.01
35	CLAYISH	WITHOUT	ABUNDANT	25	NO	NO	1.00
36	CLAYISH	WITH	VERY ABUNDANT	5	LEAVES	SNAIL	1.00
37	CLAYISH	WITH	ABUNDANT	10	NO	SNAIL	1.00
38	CLAYISH	WITH	REGULAR	15	NO	SNAIL	1.00
39	CLAYISH	WITH	ABUNDANT	10	LEAVES	SNAIL	1.00
40	CLAYISH	WITH	ABUNDANT	15	NO	NO	1.00
41	CLAYISH	WITH	ABUNDANT	25	NO	NO	1.00
42	SANDY-SILTY	LIMESTONE	NO	50	NO	NO	1.02
43	SANDY	LIMESTONE	NO	0	NO	NO	1.03
44	SANDY	LIMESTONE	NO	0	NO	NO	1.01
45	SANDY-SILTY	LIMESTONE	NO	45	NO	NO	1.00
46	SANDY-SILTY	LIMESTONE	NO	0	NO	NO	1.03
47	SILT	WITHOUT	NO	0	NO	NO	1.01

ID BAG	PERMEABILITY	SLOPE	EROSION	DRAINAGE	HUMIDITY
1	FAST	1	NO	GOOD	HUMID
2	FAST	1	NO	GOOD	HUMID
3	FAST	1	NO	GOOD	HUMID
4	FAST	1	NO	GOOD	HUMID
5	FAST	2	NO	GOOD	HUMID
6	FAST	0	NO	GOOD	HUMID
7	FAST	0	NO	GOOD	HUMID
8	SLOW	0	NO	POOR	HUMID
9	SLOW	0	NO	POOR	HUMID
10	FAST	30	NO	GOOD	HUMID
11	FAST	3	NO	GOOD	HUMID
12	FAST	3	NO	GOOD	HUMID
13	FAST	3	NO	GOOD	HUMID
14	FAST	3	NO	GOOD	HUMID
15	FAST	3	NO	GOOD	HUMID
16	FAST	3	NO	GOOD	HUMID
17	FAST	3	NO	GOOD	HUMID
18	FAST	0	NO	GOOD	HUMID
19	FAST	0	NO	GOOD	HUMID
20	FAST	0	NO	GOOD	HUMID
21	FAST	0	NO	GOOD	HUMID
22	SLOW	0	NO	POOR	HUMID
23	SLOW	0	NO	POOR	HUMID
24	SLOW	0	NO	POOR	HUMID
25	FAST	10	NO	GOOD	HUMID
26	SLOW	10	NO	POOR	HUMID
27	FAST	10	NO	GOOD	HUMID
28	FAST	0	NO	GOOD	HUMID
29	FAST	0	NO	GOOD	HUMID
30	FAST	0	NO	GOOD	HUMID
31	FAST	0	NO	GOOD	HUMID
32	FAST	1	NO	GOOD	HUMID
33	SLOW	1	NO	POOR	HUMID
34	SLOW	1	NO	POOR	HUMID
35	SLOW	1	NO	POOR	HUMID
36	FAST	4	NO	GOOD	HUMID
37	SLOW	4	NO	POOR	HUMID
38	SLOW	4	NO	POOR	HUMID
39	FAST	0	NO	GOOD	HUMID
40	SLOW	0	NO	POOR	HUMID
41	SLOW	0	NO	POOR	HUMID
42	N/A	N/A	NO	N/A	HUMID
43	N/A	N/A	NO	N/A	HUMID
44	N/A	N/A	NO	N/A	HUMID
45	N/A	N/A	NO	N/A	HUMID
46	N/A	N/A	NO	N/A	HUMID
47	N/A	N/A	no	N/A	DRY

Chemical Analyses of Calakmul Soil Profiles 2003 and 2004							
ID BAG	PIT	HORIZON	PH1	PH2	PH3	PH AVERAGE	INTERPRETATION
1	A	O	7.8	7.8	7.9	7.83	MODERATELY ALKALINE
2	A	A	8	7.9	7.9	7.93	MODERATELY ALKALINE
3	A	Bk	8.6	8.4	8.5	8.50	STRONGLY ALKALINE
4	A	Bk/C	8.7	8.6	8.8	8.70	STRONGLY ALKALINE
5	B	A	8	8.1	8.2	8.10	MODERATELY ALKALINE
6	C	O	8.1	8.2	8.2	8.17	MODERATELY ALKALINE
7	C	A	8.3	8.2	8.2	8.23	MODERATELY ALKALINE
8	C	Bk	8.3	8.4	8.6	8.43	MODERATELY ALKALINE
9	C	C	8.6	8.7	8.7	8.67	STRONGLY ALKALINE
10	D	A	7.7	7.9	7.8	7.80	MODERATELY ALKALINE
11	H	O	8.1	8.1	8.1	8.10	MODERATELY ALKALINE
12	H	A	8.3	8.4	8.2	8.30	MODERATELY ALKALINE
13	H	Bk	8.5	8.5	8.5	8.50	STRONGLY ALKALINE
14	I	O	8.2	8	8.1	8.10	MODERATELY ALKALINE
15	I	A	8.1	8	8	8.03	MODERATELY ALKALINE
16	I	Bk	8.4	8.4	8.4	8.40	MODERATELY ALKALINE
17	I	Bk/C	8.7	8.6	8.6	8.63	STRONGLY ALKALINE
18	J	O	7.7	7.7	7.7	7.70	MODERATELY ALKALINE
19	J	A	7.8	7.8	7.8	7.80	MODERATELY ALKALINE
20	J	Bk	7.9	7.9	7.8	7.87	MODERATELY ALKALINE
21	K	A	7.6	7.6	7.7	7.63	MODERATELY ALKALINE
22	K	A/Bg	7.7	7.9	7.8	7.80	MODERATELY ALKALINE
23	K	Bg	8.2	8.2	8.2	8.20	MODERATELY ALKALINE
24	K	Bg/C	8.1	8.2	8.1	8.13	MODERATELY ALKALINE
25	L	A	7.8	7.9	7.8	7.83	MODERATELY ALKALINE
26	L	Bk	7.9	8	8.1	8.00	MODERATELY ALKALINE
27	L	C	7.9	7.8	7.8	7.83	MODERATELY ALKALINE
28	M	A	8.4	8.2	8.1	8.23	MODERATELY ALKALINE
29	M	Bk	8.5	8.4	8.3	8.40	MODERATELY ALKALINE
30	N	A	8.4	8.2	8.3	8.30	MODERATELY ALKALINE
31	N	Bk	8.4	8.4	8.4	8.40	MODERATELY ALKALINE
32	O	O	7.8	7.6	7.7	7.70	MODERATELY ALKALINE
33	O	I	8.2	8.2	8.1	8.17	MODERATELY ALKALINE
34	O	III	8	7.9	7.9	7.93	MODERATELY ALKALINE
35	O	C	8.3	7.9	7.8	8.00	MODERATELY ALKALINE
36	P	I	7.6	7.6	7.6	7.60	MODERATELY ALKALINE
37	P	III	7.8	7.8	7.8	7.80	MODERATELY ALKALINE
38	P	RELL	8.3	8.5	8.5	8.43	MODERATELY ALKALINE
39	Q	O	7.5	7.3	7.2	7.33	NEUTRAL
40	Q	A	7.8	7.7	7.3	7.60	MODERATELY ALKALINE
41	Q	Bg	7.3	7.3	7.3	7.30	NEUTRAL
42	R	RELL	7.4	7.7	7.6	7.57	MODERATELY ALKALINE
43	R	RELL	7.9	7.9	7.9	7.90	MODERATELY ALKALINE
44	R	RELL	8	8.2	8.1	8.10	MODERATELY ALKALINE
45	R	RELL	7.8	7.7	7.8	7.77	MODERATELY ALKALINE
46	R	NO	12.3	12.1	12	12.13	STRONGLY ALKALINE
47	Witness	NO	9.9	10	9.9	9.93	STRONGLY ALKALINE

ID BAG	CONDUCTIVITY AVERAGE MMHOS/CM	CONDUCTIVITY INTERPRETATION	ORGANIC MATTER%	HCl AVERAGE	INTERPRETATION
1	0.31	LOW	32	3.0	CALCAREOUS
2	0.14	LOW	16	4.0	HIGHLY CALCAREOUS
3	0.09	LOW	16	4.0	HIGHLY CALCAREOUS
4	0.09	LOW	14	4.0	HIGHLY CALCAREOUS
5	0.14	LOW	18	4.0	HIGHLY CALCAREOUS
6	0.28	LOW	20	3.0	CALCAREOUS
7	0.11	LOW	12	4.0	HIGHLY CALCAREOUS
8	0.09	LOW	8	4.0	HIGHLY CALCAREOUS
9	0.07	LOW	6	4.0	HIGHLY CALCAREOUS
10	0.31	LOW	30	3.0	CALCAREOUS
11	0.12	LOW	28	4.0	HIGHLY CALCAREOUS
12	0.10	LOW	12	4.0	HIGHLY CALCAREOUS
13	0.08	LOW	14	4.0	HIGHLY CALCAREOUS
14	0.13	LOW	22	3.0	CALCAREOUS
15	0.10	LOW	18	3.0	CALCAREOUS
16	0.09	LOW	12	4.0	HIGHLY CALCAREOUS
17	0.09	LOW	14	4.0	HIGHLY CALCAREOUS
18	0.30	LOW	26	2.0	MODERADATELY CALCAREOUS
19	0.18	LOW	20	2.0	MODERADATELY CALCAREOUS
20	0.18	LOW	16	3.5	HIGHLY CALCAREOUS
21	0.23	LOW	22	0.0	NON CALCAREOUS
22	0.19	LOW	18	3.0	CALCAREOUS
23	0.19	LOW	16	4.0	HIGHLY CALCAREOUS
24	0.84	LOW	14	3.0	CALCAREOUS
25	0.18	LOW	18	3.5	HIGHLY CALCAREOUS
26	0.21	LOW	16	4.0	HIGHLY CALCAREOUS
27	5.00	HIGH	18	3.0	CALCAREOUS
28	0.13	LOW	14	4.0	HIGHLY CALCAREOUS
29	0.09	LOW	10	4.0	HIGHLY CALCAREOUS
30	0.11	LOW	20	4.0	HIGHLY CALCAREOUS
31	0.07	LOW	12	3.5	HIGHLY CALCAREOUS
32	0.24	LOW	20	3.0	CALCAREOUS
33	0.16	LOW	16	4.0	HIGHLY CALCAREOUS
34	0.17	LOW	12	4.0	HIGHLY CALCAREOUS
35	0.12	LOW	14	4.0	HIGHLY CALCAREOUS
36	0.32	LOW	22	3.0	CALCAREOUS
37	0.27	LOW	18	3.0	CALCAREOUS
38	0.21	LOW	20	3.5	HIGHLY CALCAREOUS
39	0.33	LOW	22	0.0	NON CALCAREOUS
40	0.28	LOW	20	0.0	NON CALCAREOUS
41	0.15	LOW	20	0.0	NON CALCAREOUS
42	5.00	HIGH	8	4.0	HIGHLY CALCAREOUS
43	5.00	HIGH	2	4.0	HIGHLY CALCAREOUS
44	0.56	LOW	4	4.0	HIGHLY CALCAREOUS
45	1.27	MEDIUM	18	4.0	HIGHLY CALCAREOUS
46	5.00	HIGH	14	4.0	HIGHLY CALCAREOUS
47	5.00	HIGH	22	0.5	VERY SLIGHTLY CALCAREOUS

ID BAG	AVERAGE P mg/l	INTERPRETATION P LEVEL
1	7.7	VERY LOW
2	5.5	VERY LOW
3	0	NONE
4	1.1	VERY LOW
5	6.05	VERY LOW
6	15.4	LOW
7	12.1	VERY LOW
8	27.5	MEDIUM
9	9.35	VERY LOW
10	8.25	VERY LOW
11	6.6	VERY LOW
12	7.15	VERY LOW
13	7.7	VERY LOW
14	1.65	VERY LOW
15	9.35	VERY LOW
16	4.95	VERY LOW
17	2.2	VERY LOW
18	15.4	LOW
19	23.1	MEDIUM
20	0	NONE
21	6.05	VERY LOW
22	8.8	VERY LOW
23	0	NONE
24	7.7	VERY LOW
25	7.15	VERY LOW
26	15.4	LOW
27	0	NONE
28	6.6	VERY LOW
29	3.3	VERY LOW
30	20.9	LOW
31	0	NONE
32	24.2	MEDIUM
33	20.35	LOW
34	12.65	VERY LOW
35	9.9	VERY LOW
36	9.9	VERY LOW
37	9.9	VERY LOW
38	0	NONE
39	8.8	VERY LOW
40	15.4	LOW
41	17.05	LOW
42	15.95	LOW
43	8.25	VERY LOW
44	0	NONE
45	19.25	LOW
46	8.8	VERY LOW
47	165	VERY HIGH

Soil Pits Profiles

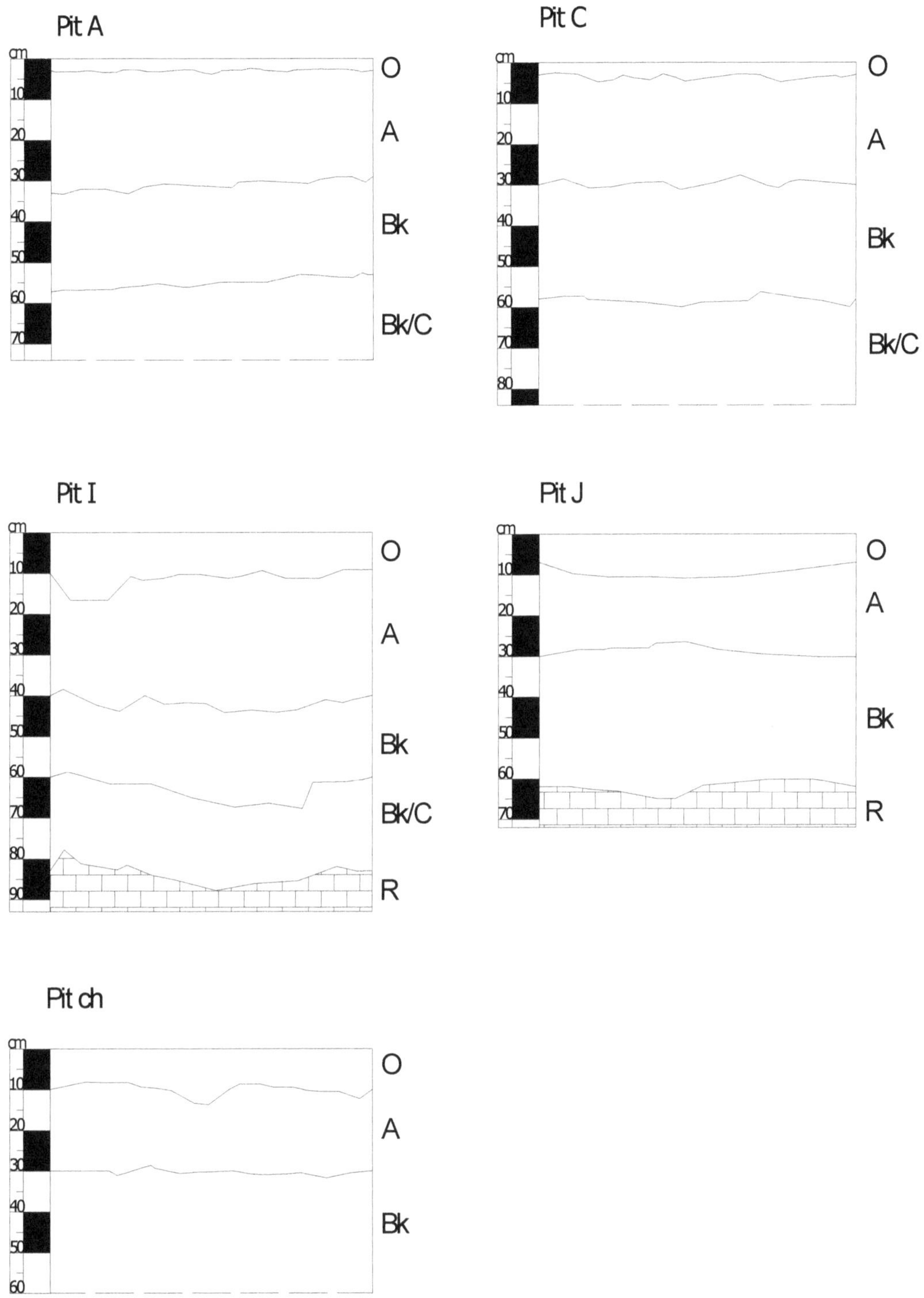

Figure A1. Soil profiles of Calcic Vertisols.

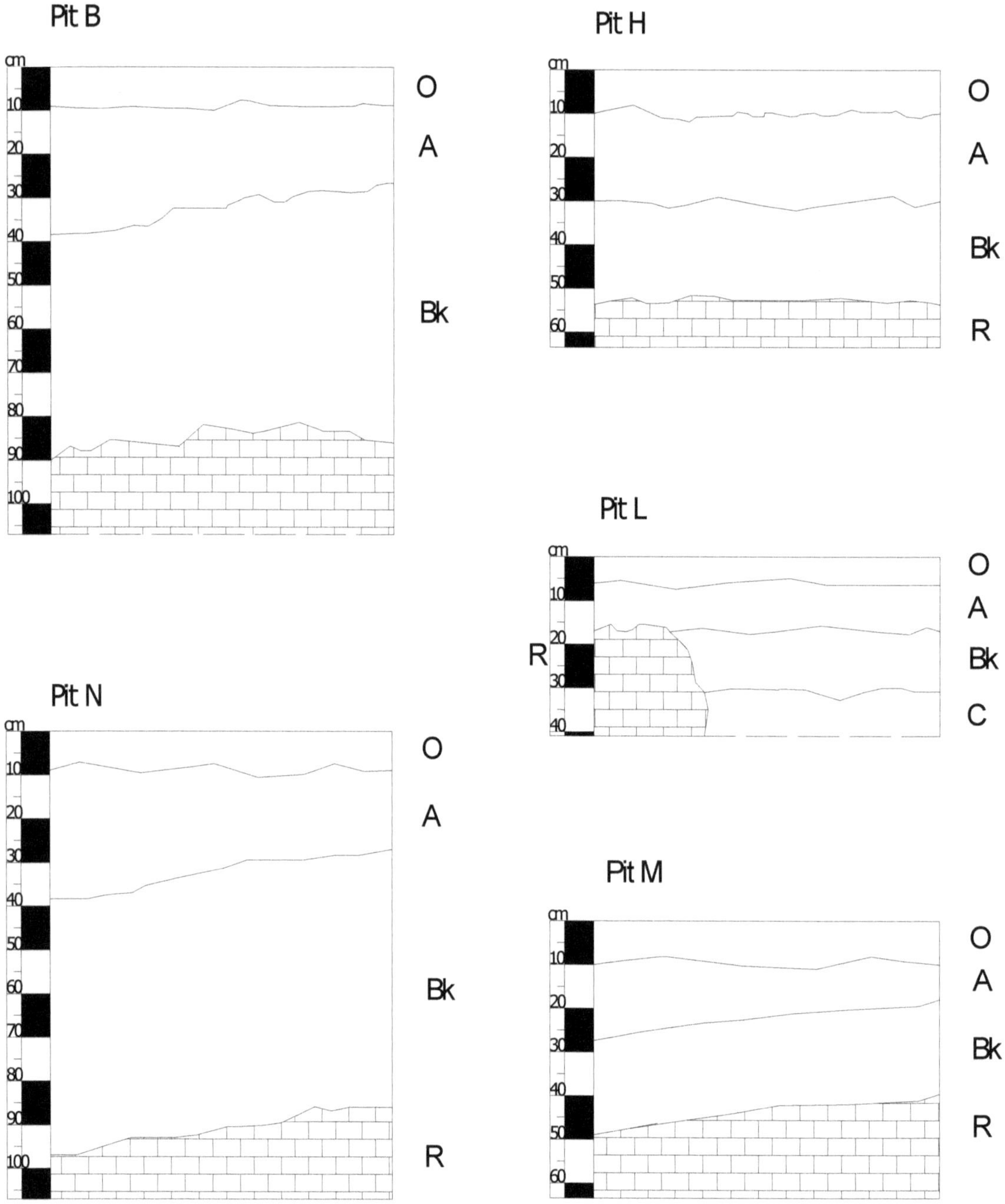

Figure A2. Soil profiles of Rendzinas.

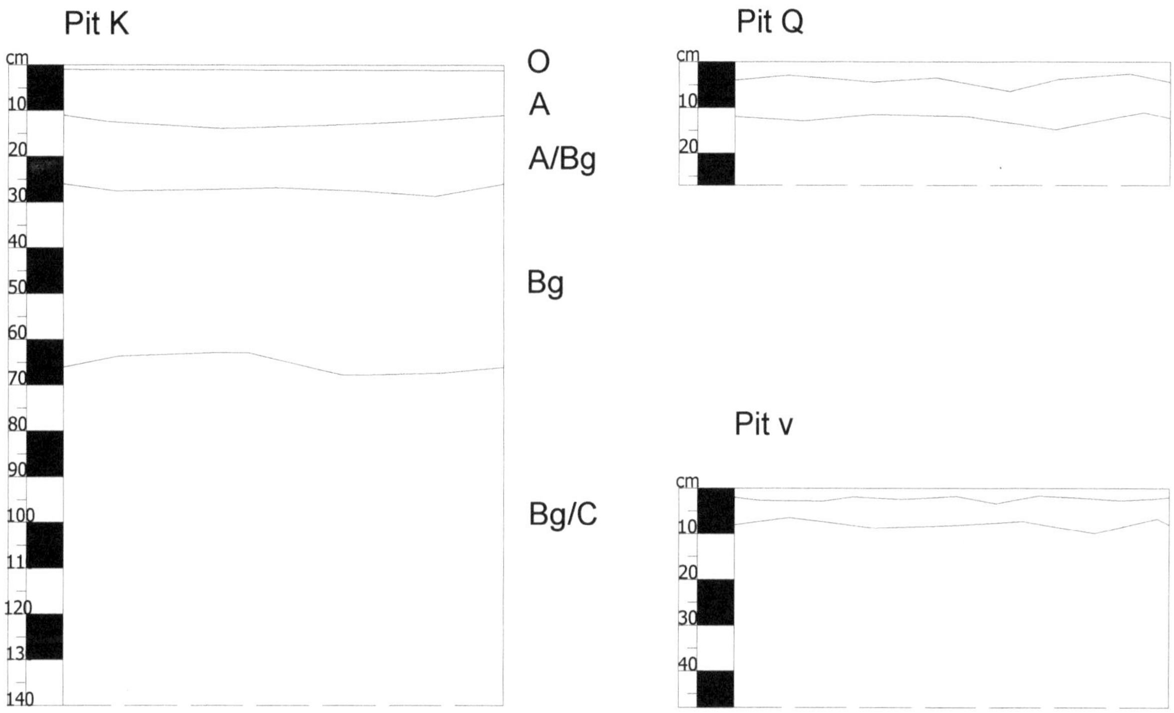

Figure A3. Soil profiles of Vertic Gleysols.

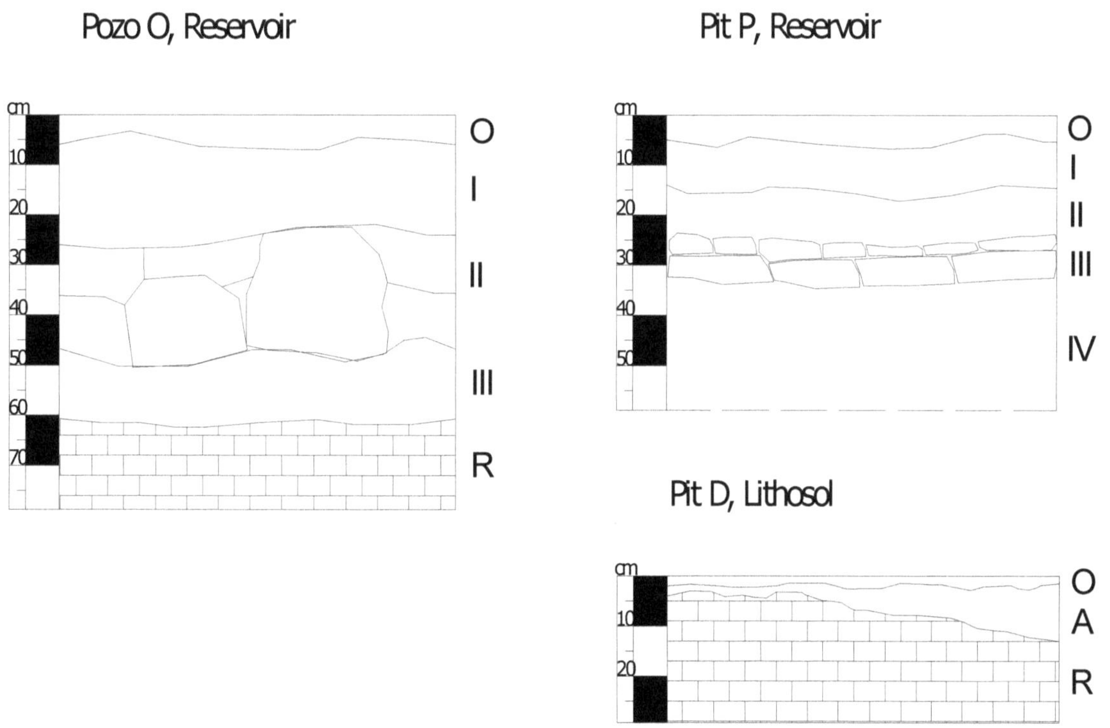

Figure A4. Soil profiles of a Lithosol and reservoirs.

APPENDIX 4. CERAMICS AND LITHICS OF EDAPHOLOGICAL PITS OF CALAKMUL

Ceramics of Edaphological Pits of Calakmul

The pits are not presented in alphabetical order. Archaeologist Alma Eugenia Martínez Dávila identified this material. I translated the results from Spanish and also modified the original format of the results.

PIT B	BAG 5				
PERIOD	**COMPLEX**	**GROUP**	**TYPE**	**VARIETY**	**HORIZON A**
TERMINAL	HALIBE	TANCACHA CAL	Otro pizarra	Non-specific	1
CLASSIC				**TOTAL**	**1**
		TRIUNFO	Triunfo estriado	Triunfo	2
		CIRICOTE	Ciricote chorreado	Ciricote	1
EARLY	KAYNIKTE	AGUILA	Aguila naranja	Aguila	18
CLASSIC		DOS ARROYOS	Yaloche crema polícromo	Yaloche	1
		MAXCANU	Tacopate chorreado/café	Tacopate	1
				TOTAL	**23**
				Without slip	1
				PIT TOTAL	**25**

PIT O	BAG 34				
PERIOD	COMPLEX	**GROUP**	**TYPE**	**VARIETY**	**LAYER I**
		TRIUNFO	Triunfo estriado	Triunfo	10
		INFIERNO	Infierno negro	Infierno	2
LATE	KU	TINAJA	Tinaja	Nanzal	50
CLASSIC		PALMAR	Huacho negro/naranja	Non-specific	3
				TOTAL	**65**
		BALANZA	Balanza negro	Balanza	3
EARLY	KAYNIKTE	AGUILA	Aguila naranja	Aguila	11
CLASSIC			Dos Arroyos polícromos	Non-specific	1
		MAXCANU	Tacopate chorreado/ café	Tacopate	1
				TOTAL	**16**
LATE	TAKAN	SIERRA	Sierra rojo	Sierra	19
PRECLASSIC				**TOTAL**	**19**
				Without slip	29
				PIT TOTAL	**129**

PIT C	BAGS 7 and 8						
PERIOD	**COMPLEX**	**GROUP**	**TYPE**	**VARIETY**	**HORIZON A**	**HORIZON Bk**	**TOTAL**
		ENCANTO	Cambio sin engobe	Non-specific		3	3
		TANCACHACAL	Pizarra tancachacal	Non-specific	1	1	2
		BALANCAN	Balancan naranja fino	Balancan	1		1
TERMINAL	HALIBE		Tenosique negro/naranja	Tenosique		1	1
CLASSIC		ALTAR	Altar naranja fino	Altar		1	1
		TINAJA	Tinaja Rojo	Tinaja	2	1	3
				TOTAL	**4**	**7**	**11**
		INFIERNO	Infierno negro	Bolocantal	2	4	6
		TINAJA	Tinaja rojo	Nanzal		19	19
			Otro Tinaja	Inciso acana		1	1
LATE	KU	CORONA	Corona rojo	Corona	1		1
CLASSIC		CHIMBOTE	Moro maranja polícromo	Non-specific		1	1
			Chimbote	Chimbote		1	1
		SAXCHE	Saxche polícromo	Saxche		1	1
		PETKANCHE	Petkanche naranja polícromo	Chorizo		1	1
				TOTAL	**3**	**28**	**31**
		TRIUNFO	Triunfo estriado	Triunfo	2	10	12
		AGUILA	Aguila	Aguila	3	7	10
			Otro Aguila	Non-specific		1	1
EARLY	KAYNIKTE		Dos Arroyos	Non-specific	7	1	8
CLASSIC		BALANZA	Balanza	Balanza	2	2	4
			Lucha inciso	Lucha		1	1
		SAMBULA	Sambula negro/gris	Non-specific		1	1
		MAXCANU	Tacopate chorreado /café	Tacopate	1	2	3
				TOTAL	**15**	**25**	**40**
LATE	TAKAN	SIERRA	Sierra rojo	Sierra		4	
PRECLASSIC				**TOTAL**		**4**	**4**
			Without slip		**3**		**3**
				TOTAL	**25**	**64**	**89**
					PIT TOTAL		**89**

PIT H	BAGS 12 AND 13						
PERIOD	**COMPLEX**	**GROUP**	**TYPE**	**VARIETY**	**HORIZON A**	**HORIZON Bk**	**TOTAL**
		ACHOTE	Achote negro	Achote	2		2
			Cubeta inciso	Non-specific	8		8
TERMINAL	HALIBE	BALANCAN	Balancan naranja fino	Balancan	4		4
CLASSIC		TINAJA	Tinaja	Tinaja	18	18	36
			Portia gubiado inciso	Portia		1	1
		PIZARRA DELGADA	Otro pizarra	Non-specific	1		1
				TOTAL	**33**	**19**	**52**
		INFIERNO	Infierno negro	Bolocantal	2	4	6
LATE	KU	CHIMBOTE	Moro naranja polícromo	Non-specific	1	1	2
CLASSIC		TINAJA	Colozal inciso	Ranurada	2		2
				TOTAL	**5**	**5**	**10**
		TRIUNFO	Triunfo estriado	Triunfo	9	4	13
			Quintal sin engobe	Quintal	5	14	19
		BALANZA	Balanza negro			5	5
EARLY	KAYNIKTE		Urita gubiado-inciso	Urita	1		1
CLASSIC			Lucha inciso			1	1
		AGUILA	Aguila naranja	Aguila	4		4
			Dos Arroyos polícromos	Non-specific	4		4
		MAXCANU	Tacopate chorreado/_café	Tacopate		5	5
				TOTAL	**23**	**29**	**52**
			Without slip		4	1	**5**
				TOTAL	**65**	**54**	**119**
						PIT TOTAL	**119**

PIT I	BAG 15				
PERIOD	**COMPLEX**	**GROUP**	**TYPE**	**VARIETY**	**HORIZON A**
		ENCANTO	Cambio sin engobe	Non-specific	4
TERMINAL	HALIBE	ACHOTE	Achote negro	Achote	2
CLASSIC		TINAJA	Tinaja	Tinaja	1
		HONTUN	Homobo acanalado	Homobo	3
				TOTAL	**10**
LATE	KU	TINAJA	Chinja impreso	Chinja	1
CLASSIC		PALMAR	Palmar naranja polícromo	Non-specific	1
				TOTAL	**2**
		TRIUNFO	Triunfo	Triunfo	3
EARLY	KAYNIKTE	BALANZA	Balanza negro	Balanza	1
CLASSIC		AGUILA	Aguila naranja	Aguila	3
				TOTAL	**7**
LATE	TAKAN	SIERRA	Sierra rojo	Sierra	1
PRECLASSIC				**TOTAL**	**1**
			Without slip		5
				PIT TOTAL	**25**

PIT L	BAG 25				
PERIOD	**COMPLEX**	**GROUP**	**TYPE**	**VARIETY**	**HORIZON A**
CLASSIC	KU	TINAJA	Tinaja	Tinaja	2
LATE				**TOTAL**	**2**
		TRIUNFO	Triunfo	Triunfo	1
EARLY	KAYNIKTE		Quintal sin engobe	Quintal	1
CLASSIC				**TOTAL**	**2**
			Without slip		2
				PIT TOTAL	**6**

PIT J	BAGS 19 AND 20						
PERIOD	**COMPLEX**	**GROUP**	**TYPE**	**VARIETY**	**HORIZON A**	**HORIZON A/BK**	**TOTAL**
TERMINAL	HALIBE	ACHOTE	Achote negro	Achote	4		4
CLASSIC		TINAJA	Tinaja rojo	Tinaja	8		8
				TOTAL	**12**		**12**
LATE	KU	ENCANTO	Encanto estriado	Encanto	9		9
CLASSIC		INFIERNO	Infierno negro	Bolocantal		2	2
				TOTAL	**9**	**2**	**11**
		TRIUNFO	Triunfo estriado	Triunfo		2	2
EARLY	KAYNIKTE	AGUILA	Aguila naranja	Aguila		5	5
CLASSIC		MAXCANU	Tacopate chorreado/ café	Tacopate		1	1
				TOTAL		**8**	**8**
LATE	TAKAN	SIERRA	Sierra rojo	Sierra		4	4
PRECLASSIC				**TOTAL**		**4**	**4**
			Without slip		4	3	7
					PIT TOTAL		**42**

PIT K	BAGS 22 AND 23						
PERIOD	**COMPLEX**	**GROUP**	**TYPE**	**VARIETY**	**HORIZON A**	**HORIZON Bg**	**TOTAL**
		ENCANTO	Cambio sin engobe	Non-specific	1		1
TERMINAL	HALIBE	PIZARRA DELGADA	Other pizarra	Non-specific	6		6
CLASSIC		TINAJA	Tinaja	Tinaja	16		16
		BALANCAN	Balancan naranja	Balancan	1		1
				TOTAL	**24**		**24**
		TINAJA	Tinaja rojo	Nanzal	13		13
LATE	KU	INFIERNO	Infierno negro	Infierno	2		2
CLASSIC		CORONA	Corona rojo	Corona	8		8
				TOTAL	**23**		**23**
		TRIUNFO	Triunfo	Triunfo	11		11
EARLY	KAYNIKTE	BALANZA	Balanza negro	Balanza	3		3
CLASSIC		AGUILA	Aguila naranja	Aguila	8		8
				TOTAL	**22**		**22**
LATE	TAKAN	SAPOTE	Sapote estriado	Sapote	6		6
PRECLASSIC		SIERRA	Sierra rojo	Sierra	3	1	4
				TOTAL	**9**		**10**
				TOTAL	**78**	**1**	**79**
					PIT TOTAL		**79**

PIT M	BAGS 28 AND 29						
PERIOD	**COMPLEX**	**GROUP**	**TYPE**	**VARIETY**	**HORIZON A**	**HORIZON Bk**	**TOTAL**
		ACHOTE	Otro grupo achote	Rojo	1		1
TERMINAL	HALIBE	TANCACHACAL	Pizarra tancachacal	Non-specific	3	4	7
CLASSIC				**TOTAL**	**4**	**4**	**8**
		ENCANTO	Cambio sin engobe	Cambio		3	3
LATE	KU	TINAJA	Tinaja	Nanzal	12	12	24
CLASSIC				**TOTAL**	**12**	**15**	**27**
		TRIUNFO	Triunfo estriado	Triunfo	4	22	26
			Quintal sin engobe	Quintal		11	11
EARLY		CIRICOTE	Ciricote chorreado	Non-specific	2		2
CLASSIC	KAYNIKTE	BALANZA	Balanza negro	Balanza	6	8	14
		AGUILA	Aguila naranja	Aguila	8	36	44
			Dos Arroyos polícromos	Non-specific		1	1
		MAXCANU	Tacopate chorreado/_ café	Tacopate		20	20
				TOTAL	**20**	**98**	**118**
		SAPOTE	Sapote estriado	Sapote	9	28	37
LATE	TAKAN	SIERRA	Sierra rojo	Sierra	12	8	28
PRECLASSIC			Laguna verde inciso		1		1
		POLVERO	Polvero negro			4	4
				TOTAL	**22**	**40**	**62**
			Without slip		1	1	2
			Figurine		1		1
				TOTAL	**60**	**158**	**218**
					PIT TOTAL		**218**

PIT N	BAGS 30 AND 31						
PERIOD	**COMPLEX**	**GROUP**	**TYPE**	**VARIETY**	**HORIZON A**	**HORIZON B/K**	**TOTAL**
		ACHOTE	Achote negro	Achote	9	9	18
		PIZARRA DELGADA	Pizarra delgada	Non-specific		3	3
		TANCACHACAL	Pizarra tancachacal	Non-specific		9	9
TERMINAL	HALIBE		Pizarra incisa	Incisa		1	1
CLASSIC		BALANCAN	Balancan naranja fino	Balancan	1	3	4
		TINAJA	Tinaja	tinaja	32	7	39
			Pantano impreso	Pantano	1		1
				TOTAL	**43**	**32**	**75**
		INFIERNO	Infierno negro	Bolocantal	3	9	12
		TINAJA	Tinaja	Nanzal	28	15	43
LATE	KU		Corozal inciso	Non-specific	1		1
CLASSIC		SAXCHE	Saxche naranja polícroma	Saxche		3	3
		TRAINO	Traino café	Traino		1	1
				TOTAL	**32**	**28**	**60**
		TRIUNFO	Triunfo estriado	Triunfo	16	4	20
			Quintal sin engobe	Quintal	13	6	19
		CIRICOTE	Ciricote chorreado		9	8	17
EARLY	KAYNIKTE	BALANZA	Balanza negro	Balanza	3	5	8
CLASSIC			Lucha inciso	Lucha	3		3
		AGUILA	Aguila naranja	Aguila	8	13	21
			Dos Arroyos polícromos	Non-specific	1		1
			Boleto negro/naranja	Non-specific		1	1
		PUCTE	Pucte café	Non-specific		7	7
				TOTAL	**53**	**44**	**97**
		SAPOTE	Sapote estriado	Sapote	5	22	27
		SIERRA	Sierra rojo	Sierra	4	21	25
			Repasto negro/rojo	Repasto	1	2	3
LATE	TAKAN		Laguna verde inciso	Laguna verde		2	2
PRECLASSIC		POLVERO	Polvero negro	Polvero		2	2
		FLOR CREMA	Mateo rojo/crema	Non-specific		5	5
				TOTAL	**10**	**54**	**64**
			Possible figurine	Figurine		1	1
				Without slip	1	7	8
				TOTAL	**139**	**166**	**305**
					PIT TOTAL		**305**

Lithics of Edaphological Pits of Calakmul

FORMER ID	ID BAG	PIT	LAYER	QUANTITY	DESCRIPTION
H4	12	H	A	3	LIMESTONE WITH FACES
H5	13	H	Bk	1	LIMESTONE TOOL PREPARED FOR HAFTING IN A WOODEN HANDLE
				1	FLINT FRAGMENT
H6	15	I	A	3	FLINT FRAGMENTS
H7	19	J	A	2	FLINT FRAGMENTS
H8	20	J	Bk	1	LIMESTONE PICK PREPARED FOR HAFTING IN A WOODEN HANDLE
H10	23	K	Bg	3	FLINT FRAGMENTS
H11	30	N	A	4	FLINT FRAGMENTS
H13	25	L	A	1	LIMESTONE WITH HANDLE TRACE
				4	FLINT FRAGMENTS
H14	28	M	A	1	LIMESTONE FRAGMENT
				6	FLINT FRAGMENTS
H15	29	M	Bk	4	FLINT FRAGMENTS
				2	FLINT WITH CORTEX
H16	33	O	RELL	1	LIMESTONE ROCK
				11	FLINT FRAGMENTS
				1	FLINT FRAGMENT
				1	GRINDING STONE (CRISTAL ROCK) WITH RED PIGMENT
			TOTAL	50	

REFERENCES

Acopa, D. and E. Boege
1998 The Maya Forest in Campeche, Mexico: Experiences in Forest Management at Calakmul. In *Timber, Tourists and Temples*, edited by R. B. Primack, D. B. Bray, H. A. Galletti and I. Ponciano. pp: 81-97 Island Press, Washington, D. C.

Adams, R. E. W.
1986 Rio Azul. *National Geographic* 169(4):420-451.

1990 Archaeological Investigations at the Lowland Maya Site of Rio Azul. *Latin American Antiquity* 1(1):23-41.

Adams, R. E. W., W. E. Brown Jr. and T. P. Culbert
1981 Radar Mapping, Archaeology and Ancient Maya Land Use. *Science* 213:1457-1463.

Adams, R. E. W. and R. C. Jones
1981 Spatial Patterns and Regional Growth among Classic Maya Cities. *American Antiquity* 46(2):301-322.

Aguilar, A.
1987 *Análisis Químico para Evaluar la Fertilidad del Suelo*. Sociedad Mexicana de la Ciencia del Suelo, México, D. F.

Aliphat, M. M.
1994 *Classic Maya Landscape in the Upper Usumacinta River Valley.* Doctoral dissertation, University of Calgary.

Aliphat, M. M., M. Zolá and H. Geovannini
1997 Edafología de la Cuenca del Río Lacanhá, Chiapas, México. Preliminary Study. Report in File. Colegio de Posgraduados, Puebla.

Arias, L. M.
1994 La Producción Milpera Actual en Yaxcabá, Yucatán. In *La Milpa en Yucatán. Un Sistema de Producción Agrícola Tradicional*, edited by H. X, B. E. Bello and T. S. Levy, pp. 171-197. Colegio de Posgraduados, Montecillos.

Ashton, M. and T. Rowley
1974 *Landscape Archaeology*. David and Charles, Newton Abbot.

Barnhart, E.
1999 *The Palenque Mapping Project, 1999 Field Season Report*. Report submitted to Foundation for Advancement of Mesoamerica, http://www.famsi.org/reports/98063/index.html.

Barrera Vázquez, A. A.
1995 *Diccionario Maya. Maya-Español, Español-Maya*. Third ed. Editorial Porrúa, México, D. F.

Barrera Vázquez, A. A., A. Gómez Pompa and C. Vázquez Yanes
1977 El Manejo de las Selvas por los Mayas: Sus Implicaciones Silvícolas y Agrícolas. *Biótica* 2(2):47-61.

Bastarrachea, J. R., E. Yah Pech and F. Briceño Chel
1992 *Diccionario Básico Español/ Maya/ Español*. Biblioteca Básica del Mayab. Maldonado Editores, Mérida.

Bautista, F., D. Palma-López and W. Huchin-Malta
2005 Actualización de la Clasificación de los Suelos del Estado de Yucatán. In *Caracterización y Manejo de los Suelos de la Península de Yucatán: Implicaciones Agropecuarias, Forestales y Ambientales*, edited by F. Bautista and G. Palacio, pp. 105- 122. Universidad Autónoma de Campeche and Universidad Autónoma de Yucatán, Mérida.

Bautista, F., A. Zinck and H. Estrada Medina
2006 Mayan Soil Classification in the Yucatan State, Mexico. Paper presented at the 18th World Congress of Soil Science, Philadelphia.

Beach, T., S. Luzzadder Beach, N. P. Dunning, J. Hageman and J. Lohse
2002 Upland Agriculture in the Maya Lowlands: Ancient Maya Soil Conservation in Northwestern Belize. *Geographical Review*: 372-397.

Beach, T., N. P. Dunning, S. Luzzadder Beach and V. Scarborough
2003 Depression Soils in the Lowland Tropics of Northwestern Belize: Anthropogenic and Natural Origins. In *Lowland Maya Area: Three Millennia at the Human-Wildland Interface*, edited by A. Gómez-Pompa, M. F. Allen and S. L. Fedick, pp 139-174. Food Products Press, Binghamton.

Bell, M. and M. J. C. Walker
1992 *Late Quaternary Environmental Change*. Longman Scientific and Technical, Essex.

Bennet, J. W.
1976 *The Ecological Transition: Cultural Anthropology and Human Adaptation.* Pergamon Press, Elmsford.

Berlin, H.
1958 El Glifo "Emblema" en las Inscripciones Mayas. *Journal de la Société des Américanistes* (47):111-119.

Binford, L. R.
2001 *Constructing Frames of Reference: An Analytical Method for Archaeological Theory Building Using Hunter-Gatherer and*

Environmental Data Sets, University of California Press, Berkeley.

Binford, M. W., M. Brenner, T. J. Whitmore, A. Higuera Gundy and S. D. Deevey
1987 Ecosystems, Paleoecology and Human Disturbance in Subtropical and Tropical America. *Quaternary Science Review* 6:115-128.

Bloom, P. R., M. D. Pohl, C. Buttleman, F. Wiseman, A. Covich, C. Miksicek, J. Ball and J. Stein
1983 Prehistoric Maya Wetland Agriculture and the Alluvial Soils near San Antonio Rio Hondo, Belize. *Science* 301(3):417-419.

Boege, E.
1993 El Desarrollo Sustentable y la Reserva de la Biosfera de Calakmul, Campeche, México. *Boletín de Antropología Americana* 28:99-132.

Boucher, S. and Y. Palomo
2000 Cerámica Ritual de Calakmul. *Arqueología Mexicana* 7(42):34-39.

Boucher, S. and N. Rojas
1995 Resultados Preliminares del Análisis Cerámico del Proyecto Arqueológico de la Biosfera de Calakmul. Paper presented at the 1st International Symposium of Maya Archaeology, San Ignacio, Belize.

Bradford, J.
1957 *Ancient Landscapes: Studies in Field Archaeology*. Bell, London.

Braswell, G. E. (editor)
2003 *The Maya and Teotihuacan: Reinterpreting Early Classic Interaction*. 1st ed. University of Texas Press, Austin.

Braswell, G., J. D. Gunn, M. d. R. Domínguez Carrasco, W. J. Folan, L. A. Fletcher, A. Morales López and M. D. Glascock
2004 Defining the Terminal Classic at Calakmul, Campeche. In *The Terminal Classic in the Maya Lowlands: Collapse, Tradition and Transformation*, edited by A. A. Demarest, P. M. Rice and D. S. Rice, pp. 102-124. University Press of Colorado,Boulder.

Breimer, R. F., A. J. van Kekem and H. van Reuler
1986 *Guidelines for Soil Survey and Land Evaluation in Ecological Research*. Man and Biosphere Technical Notes 17. United Nations Educational, Scientific and Cultural Organization, Paris.

Brown, J. C.
2005 *A Study of Forest Communities and Woody Plant Distributions in the Calakmul Biosphere Reserve, Campeche, Mexico*. PhD Dissertation, Tulane University.

Butterlin, J. and F. Bonet
1963 *Las Formaciones Cenozoicas de la Península de Yucatán. Ingeniería Hidráulica en México*. Instituto de Geología, Universidad Nacional Autónoma de México, México, D. F.

Butzer, K. W.
1982 *Archaeology as Human Ecology: Method and Theory for a Contextual Approach*. Cambridge University Press, Cambridge.

Carmack, R. M.
1981 *The Quiché Mayas of Utatlán : The Evolution of a Highland Guatemala Kingdom*. 1st ed. The Civilization of the American Indian Series 155. University of Oklahoma Press, Norman.

Carrasco, R.
2000 El Chuchcabal de la Cabeza de Serpiente. *Arqueología Mexicana* 7(42):12-19.

2003 *Calakmul en el Arte y la Arquitectura del Formativo*. Calakmul. Antología. Gobierno del Estado de Campeche. Instituto de Cultura. Universidad Autónoma de Campeche, Campeche.

Carrasco, R., A. Anaya, O. Rodríguez, R. García and C. García
2000 Proyecto Arqueológico Calakmul. Informe Temporada 1999. Manuscript in Archivo Técnico, Instituto Nacional de Antropología e Historia, México, D. F.

Carrasco, R. and A. Bojalil
2005 Nuevos Datos para la Historia del Arte y la Iconografía del Clásico Temprano en el Area Maya: el Reino de Ka´an. *Boletín Informativo. La Pintura Mural Prehispánica en México* 23(11):24-32.

Carrasco, R. y S. Boucher
1994 Espacios Sagrados y Objetos de Poder. *Arqueología Mexicana* 2(10):32-38.

Carrasco, R., S. Boucher, P. Alvarez, V. Tiesler, V. García, R. García and J. Vázquez
1999b A Dynastic Tomb from Campeche, Mexico: New Evidence on Jaguar Paw, a Ruler of Calakmul. *Latin American Antiquity* 10(1):47-58.

Carrasco, R., S. Boucher, C. Baudez, J. P. Courau, A. Dowd, M. Wolf, E. Gonzalez, S. Martin, V. García, R. Schneider, G. Sánchez, R. Carrillo, A. Paul, N. Rojas and M. Brook

1996 Informe Técnico Final Proyecto Arqueológico de la Biósfera Calakmul, Campeche. Manuscript in Archivo Técnico, Instituto Nacional de Antropología e Historia, México, D. F.

Carrasco, R., S. Boucher, A. Cantero, P. Alvarez, E. Gonzalez, V. Tiesler, C. García, R. García, J. Granados, S. Dzul and O. Rodríguez
1999a Informe de los Trabajos Arqueológicos Calakmul, Temporada 1998 –1999. Manuscript in Archivo Técnico, Instituto Nacional de Antropología e Historia, México, D.F.

Carrasco, R., S. Boucher, E. Gonzalez, Y. Palomo, P. Alvarez, L. E. Campaña, V. Tiesler, D. Reents Budet, R. Bishop, V. García, J. Granados, R. García, S. Dzul, S. Martin and J. Vazquez
1998 Informe de los Trabajos Arqueológicos de Calakmul. Temporada 1997 -1998. Manuscript in Archivo Técnico, Instituto Nacional de Antropología e Historia, México, D. F.

Carrasco, R. and M. Colón
2005 El Reino de Kaan y la Antigua Ciudad de Calakmul. *Arqueología Mexicana* 13(75):40-47.

Carrasco, R., O. Rodríguez and A. Anaya
2002 Proyecto Arqueológico Calakmul. Infome Temporada 2001. Manuscript in Archivo Técnico, Instituto Nacional de Antropología e Historia, México, D. F.

Carrasco, R., O. Rodríguez, A. Anaya, S. Boucher, Y. Palomo, S. Dzul, B. Curiel, R. García, M. Colón and K. Delvendahl
2001 Proyecto Arqueológico Calakmul. Informe Temporada 2000. Manuscript in Archivo Técnico, Instituto Nacional de Antropología e Historia, México, D. F.

Carter, W. E.
1969 *New Lands and Old Traditions: Kekchi Cultivators in the Guatemalan Lowlands.* Latin American Monograph No. 6. University of Florida Press, Gainesville.

Ceja Tenorio, J. F.
1985 *Paso de la Amada: An Early Preclassic Site in the Soconusco, Chiapas, Mexico.* Papers of the New World Archaeological Foundation, No. 49. Brigham Young University, Provo.

Centro Agronómico de Investigación y Enseñanza (CATIE)
1990 *Guía par el Manejo Integrado del Cutivo de Maíz.* Proyecto Regional Manejo Integrado de Plagas. CATIE, Turrialba.

Clark, C. D., S. M. Garrod and M. Parker Pearson
1998 Landscape Archaeology and Remote Sensing in Southern Madagascar. *Remote Sensing* 19(8):1461-1477.

Clark, G.
1964 *Archaeology and Society. Reconstructing the Prehistoric Past.* Methuen and Co., New Fetter Lane.

Coe, M. D.
2000a *The Maya.* Sixth ed. Thames & Hudson, London.

2000b *Mexico. From the Olmecs to the Aztecs.* Fourth ed. Thames & Hudson, London.

Colunga García Marín, P. and D. Zizumbo Villareal
2004 Domestication of Plants in Maya Lowlands. *Economic Botany* 58:101-110.

Comisión Nacional del Agua
1997 *Diagnóstico de la Región XII, Península de Yucatán.* Gerencia Regional de la Península de Yucatán, Mérida.

2005 *Registros de Temperatura, Precipitación y Evaporación de las Estaciones Noh Yaxche y Laguna de Alvarado, Campeche, México.* Gerencia Regional de la Península de Yucatán, Mérida.

2007 *Registros de Lluvia, Humedad y Temperatura de la Estación Meteorológica Automática Calakmul.* Gerencia Regional de la Península de Yucatán, Mérida.

Consejo Nacional de Población (CONAPO)
2006 *Población Total 2006.* Dirección de Planeación y Desarrollo, México, D.F

Cook, S.
1972 Prehistoric Demography. Addison-Wesley Modular Publication 16. Addison-Wesley Publishing, Massachusetts.

Cooke, C. W.
1931 Why the Mayan Cities of the Peten District, Guatemala, Were Abandoned. *Journal of the Washington Academy of Sciences* 21(13):283-287.

Cowgill, U. M.
1961 Soil Fertility and The Ancient Maya. *Transactions of the Conneticut Academy of Art and Sciences* 42:1-56.

Criado, F.
1997 Introduction: Combining the Different Dimensions of Cultural Space: Is a Total Archaeology of Landscape Possible? In *Landscape, Archaeology, Heritage*, edited by F. Criado and C. Parcero, pp. 5-9. Grupo de

Investigación en Arqueología del Paisaje 2. Universidade de Santiago de Compostela, Santiago de Compostela.

1999 *Del Terreno al Espacio: Planteamientos y Perspectivas para la Arqueología del Paisaje.* Grupo de Investigación en Arqueología del Paisaje 6. Universidade de Santiago de Compostela, Santiago de Compostela.

Crumley, C. L.
1987 Historical Ecology. In *Regional Dynamics: Burgundian Landscapes in Historical Perspective*, edited by C. L. Crumley and W. H. Marquardt, 237-264. Academic Press, San Diego.

Culbert, T. P.
1995 Ancient Maya Wetland Agriculture. Report Submitted to Foundation for the Advancement of Mesoamerican Studies, http://www.famsi.org/reports/94033.

Culbert, T. P., L. J. Levy and L. Cruz
1990 Lowland Maya Wetland Agriculture. In *Vision and Revision in Maya Studies*, edited by F. Clancy and P. D. Harrison, pp. 115-124. University of New Mexico Press, Albuquerque.

Culbert, T. P. and D. S. Rice (editors)
1990 *Precolumbian Population History in the Maya Lowlands*. University of New Mexico Press, Albuquerque.

Curtis, J. H., M. Brenner, D. A. Hodell, R. A. Balser, G. A. Islebe and H. Hooghiemstra
1998 A Multi-proxy Study of Holocene Environmental Change in the Maya Lowlands of Peten, Guatemala. *Journal of Paleolimnology* 19:139-159.

Curtis, J. H., D. A. Hodell and M. Brenner
1996 Climate Variability on the Yucatan Peninsula (Mexico) During the past 3500 Years, and Implications for Maya Cultural Evolution. *Quaternary Research* 46:37-47.

Dahlin, B.
1985 La Geografía Histórica de la Antigua Cultura Maya. In *Historia de la Agricultura. Época Prehispánica S. XVI*, edited by T. Rojas and W. Sanders, pp. 125-196. Col. Biblioteca del Instituto Nacional de Antropología e Historia, México, D. F.

Dahlin, B., Beach, T., Luzzadder-Beach, S., Hixson, D. and Hutson, S.
2005 Reconstructing Agricultural Self-Sufficiency at Chunchucmil, Yucatan, Mexico. *Ancient Mesoamerica* 16:229-247.

Davis, D. E.
2003 *GIS for Everyone*. ESRI Press, Redlands.

Deevey, E. S., D. S. Rice, P. M. Rice, H. H. Vaughan, M. Brenner and M. S. Flannery
1979 Mayan Urbanism: Impact on a Tropical Karst Environment. *Science* 206:298-306.

Demarest, A. A.
2006 *Ancient Maya*. Cambridge University Press, Cambridge.

de Landa, F. D.
1994 *Relación de las Cosas de Yucatán*. Consejo Nacional para la Cultura y las Artes , México, D. F.

Dent, D. and A. Young
1981 *Soil Survey and Land Evaluation*. George Allen and Unwin, London.

Díaz Gallegos, J. R., O. Castillo-Acosta and G. García Gil
2001 Distribución Espacial y Estructura Arbórea de la Selva Baja Subperennifolia en un Ejido de la Reserva de la Biosfera Calakmul. *Universidad y Ciencia* 18(34):11-28.

Díaz Ponce, A.
1999 Regiones Hidrológicas, Escala 1:4 000 000. Comisión Nacional del Agua.

Domínguez Carrasco, M. d. R.
1994 *Calakmul, Campeche: Un Análisis de la Cerámica.* Colección Arqueología 4. Centro de Investigaciones Históricas y Sociales, Universidad Autónoma de Campeche, Campeche.

1996 La Cerámica de Calakmul, Campeche: Una Visión de su Secuencia Cronológica y Cultural. In *Los Investigadores de la Cultura Maya*, pp. 504-521. vol. 3, tomo II. Universidad Autónoma de Campeche, Campeche.

Domínguez Carrasco, M. d. R. and W. J. Folan
1996 Calakmul, México: Aguadas, Bajos, Precipitación y Asentamiento en el Petén Campechano. In *IX Simposio de Investigaciones Arqueológicas en Guatemala*, edited by J. P. Laporte and H. I. Escobedo, pp. 171-193. Museo Nacional de Arqueología y Etnología, Guatemala.

Drennan, R. D.
1988 Household Location and Compact Versus Dispersed Settlement in Prehispanic Mesoamerica. In *Household and Community in the Mesoamerican Past*, edited by R. R. Wilk

and W. Ashmore, pp. 273-293. University of New Mexico Press, Albuquerque.

Duch, G. J.
1989 *Los Bajos Inundables (Akálche') de la Península de Yucatán.* Universidad Autónoma de Chapingo, México, D. F.

Dunning, N. P.
1992 *Lords of the Hills: Ancient Maya Settlement in the Puuc Region, Yucatán, México.* Monographs in World Archaeology No. 15. Prehistory Press, Madison.

1996 A Reexamination of Regional Variability in the Prehistoric Agricultural Landscape. In *The Managed Mosaic: Ancient Maya Agriculture,* edited by S. L. Fedick, pp. 53-68. University of Utah Press, Salt Lake City.

Dunning, N. P. and T. Beach
2000 Stability and Instability in Prehispanic Maya Landscape. In *Imperfect Balance: Landscape Transformations in the Precolumbian Americas,* edited by D. L. Lentz, pp. 179-202. Columbia University Press, New York.

Dunning, N. P., J. G. Jones, T. Beach and S. Luzzadder Beach
1999 Temple Mountains, Sacred Lakes, and Fertile Fields: Ancient Maya Landscapes in Northwestern Belize. *Antiquity* 73:650-660.

Dunning, N. P., S. Luzzadder Beach, T. Beach, J. G. Jones, V. Scarborough and T. P. Culbert
2002 Arising from the Bajos: The Evolution of a Neotropical Landscape and the Rise of Maya Civilization. *Annals of the Association of American Geographers* 92(2):267-283.

Erickson, C. L.
1988 Raised Field Agriculture in the Lake Titicaca Basin: Putting Ancient Andean Agriculture Back to Work. *Expedition* 30(3):8-16.

1999 Neo-environmental Determinism and Agrarian "Collapse" in Andean Prehistory. *Antiquity* 73:634-642.

2001 Pre-Columbian Roads of the Amazon. *Expedition* 43(2):21-30.

2003 Historical Ecology and Future Explorations. In *Amazonian Dark Earths: Origins, Properties, Management,* edited by J. Lehmann, D. Kern, B. Glaser and W. Woods, pp. 455-500. Kluwer, Dordrecht.

Escamilla, A., M. Sanvincente, M. Sosa and C. Galindo-Leal
2000 Habitat Mosaic, Wildlife Availability, and Hunting in the Tropical Forest of Calakmul, Mexico. *Conservation Biology* 14(6):1592-1601.

Espinosa, M., V. Rodríguez, M. d. R. Domínguez Carrasco, W. J. Folan and H. Chung
2001 Estudio Preliminar de las Herramientas de Pedernal Procedentes de Calakmul a Través de Técnicas Analíticas. In *Los Investigadores de la Cultura Maya,* pp. 268-283. vol. 9. tomo II. Universidad Autónoma de Campeche, Campeche.

Eyhorn, F., M. Heeb and G. Weidmann
2002 *IFOAM Training Manual for Organic Agriculture in the Tropics. Theory, Transparencies, Didactic Approach.* International Federation of Organic Agriculture Movements (IFOAM), Tholey-Theley.

Fash, W. L.
1991 *Scribes, Warriors, and Kings.* Thames and Hudson, London.

Fedick, S. L.
1994 Ancient Maya Agricultural Terracing in the Upper Belize River Area. *Ancient Mesoamerica* 5:107-127.

1995 Land Evaluation and Ancient Maya Land Use in the Upper Belize River Area, Belize, Central America. *Latin American Antiquity* 6(1):16-34.

1996 Introduction: New Perspectives on Ancient Maya Agriculture and Resource Use. In *The Managed Mosaic: Ancient Maya Agriculture,* edited by S. L. Fedick, pp. 1-14. University of Utah Press, Salt Lake City.

Fedick, S. L. and B. A. Morrison
2004 Ancient Use and Manipulation of Landscape in the Yalahau Region of the Northern Maya Lowlands. *Agriculture and Human Values* 21:207–219.

Flannery, K. V.
1972 The Cultural Evolution of Civilizations. *Annual Review of Ecology and Systematics* 3(3):399-426.

Fletcher, L. and J. Gann
1987 Calakmul, Campeche: Patrón de Asentamiento y Demografía. In *Seis Ensayos Sobre Antiguos Patrones de Asentamiento en el Area Maya,* edited by E. Vargas Pacheco, pp 53-67. Instituto de Investigaciones Antropológicas, Universidad Nacional Autónoma de México, México, D. F.

Flores, J. S.

1990 Reporte Preliminar de la Vegetación de la Reserva de la Biosfera Calakmul, Campeche. Facultad de Medicina Veterinaria y Zootecnia. Universidad Autónoma de Yucatán, Mérida.

Flores-Díaz, A.
1975 *Uso de Suelo y Fertilizantes en la Epoca Prehispánica.* Cuaderno de Trabajo INAH, No. 9, pp. 51-66. Instituto Nacional de Antropología e Historia, México, D. F.

Flores-Nava, A.
1994 Some Limnological Data From Five Water Bodies of Yucatan as a Basis for Aquaculture Development. *Anales del Instituto de Ciencias del Mar y Limnología* 21: 1-2

Folan, W. J.
1985 Informe Preliminar del Proyecto Calakmul 1984-1985. Manuscript in Archivo Técnico, Instituto Nacional de Antropología e Historia, México, D. F.

Folan, W. J., M. d. R. Domínguez Carrasco, A. Morales López, R. González Heredia, J. A. Hernández, L. Florey Folan, M. Coyoc Ramírez and J. D. Gunn
2003 Calakmul, Campeche: La Ciudad Prehispánica y Estado Regional, la Reserva de la Biosfera, el Municipio Ecológico y la Declaratoria como Patrimonio Cultural de la Humanidad: Veintidos Años de Trabajo Científico de la Universidad Autónoma de Campeche y su Centro de Investigaciones Históricas y Sociales. *Calakmul. Antología*, pp. 111-126. Gobierno del Estado de Campeche. Instituto de Cultura. Universidad Autónoma de Campeche, Campeche.

Folan, W. J. and S. Gallegos
1998 Uso de Suelo en el Estado de Campeche, México y Alrededores. In *Los Investigadores de la Cultura Maya*, pp. 459-478. vol. 2. Universidad Autónoma de Campeche, Campeche.

Folan, W. J. and J. M. García Ortega
2000 Reserva de la Biosfera de Calakmul: Los Primeros Esfuerzos. Paper presented at the XIV Simposio de Investigaciones Arqueológicas, Guatemala.

Folan, W. J., J. M. García Ortega and M. Sánchez
1999 *Naturaleza y Cultura en Calakmul, Campeche.* Centro de Investigaciones Históricas y Sociales, Universidad Autónoma de Campeche and SEP, Campeche.

Folan, W. J., L. A. Fletcher, J. May Hau and L. Florey Folan
2001a *Las Ruinas de Calakmul, Campeche, México: Un Lugar Central y su Paisaje Cultural.* Centro de Investigaciones Históricas y Sociales, Universidad Autónoma de Campeche, Secretaría de Educación Pública, FAMSI & FOMES, Campeche.

Folan, W. J., E. R. Kintz and L. Fletcher
1983 *Coba: A Classic Maya Metropolis.* Academic Press, New York.

Folan, W. J., J. Marcus, S. Pincemin, M. d. R. Domínguez Carrasco, L. A. Fletcher and A. Morales López
1995 Calakmul, Campeche: New Data from an Ancient Maya Capital in Campeche, Mexico. *Latin American Antiquity* 6(4):310-334.

Folan, W. J., J. May Hau, J. Marcus, W. F. Miller and R. Gonzalez Heredia
2001b Los Caminos de Calakmul, Campeche. *Ancient Mesoamerica* 12:293-298.

Food and Agriculture Organization of the United Nations (FAO)
1970 *El Reconocimiento de los Suelos de la Península de Yucatán.* Informe Técnico 1. FAO, Rome.
1974 *Soil Map of the World (1:5 000 000).* Legend. UNESCO, Paris.

1976 *A Framework for Land Evaluation.* FAO Soils Bulletin 32. FAO, Rome.

1989 Soil Conservation for the Small Farmers in the Humid Tropics. *FAO Soil Bulletin* 60:1-99.

1992 FAO-UNESCO Soil Map of the World Digital Version. FAO and UNESCO, Rome.

1993 *Guidelines for Land-Use Planning.* FAO Development Series 1. FAO, Rome.

1999 Land and Crop Management on the Hilly Terrains of Central America. *FAO Soil Bulletin* 76:1-72.

2003 Map of World Soil Resources. FAO, EC and ISRIC, Rome.

Ford, A.
2003 *Crecimiento de Población y Complejidad Social: Un Examen de Asentamiento y Medio Ambiente en las Tierras Bajas Mayas Centrales.* Serie Monográfica 14. Centro de Investigaciones Regionales de Mesoamérica y Plumsock Mesoamerican Studies, La Antigua.

Foth, H. D.
1970 *A Study of Soil Science.* 2nd ed. LaMotte Company, Chestertown.

Freidel, D. E.

1982 Subsistence, Trade and Development of the Coastal Maya. In *Maya Subsistence: Studies in Memory of Dennis E. Puleston*, edited by K. V. Flannery, pp. 131-155. Academic Press, New York.

French, C. A. I.
1994 Soils in Archaeology: Landscape Evolution and Human Occupation. *Antiquity* 68(259):469.

Freter, A.
1994 The Classic Maya Collapse at Copan, Honduras: An Analysis of Maya Rural Settlement Trends. In *Village Communities in Early Complex Societies*, edited by G. M. Schwartz and S. E. Falconer, pp. 160–176. Smithsonian Institution Press, Washington, D. C.

Gallegos, J.
1985 Excavación de las Canteras de la Temporada 1985 de Calakmul. Manuscript in Archivo Técnico, Instituto Nacional de Antropología e Historia, México, D. F.

1994 Explotación de Piedra Caliza en el Petén Campechano. *Mayab* 9:8-17.

Gallegos, J., R. Piña Chan and W. J. Folan
1985 Excavación en la Estructura VII de Calakmul Campeche. Temporada Septiembre-Diciembre 1984; Enero-Mayo 1985. Manuscript in Archivo Técnico, Instituto Nacional de Antropología e Historia, México, D. F.

Garcia Bárcena, J.
1979 *Una Punta Acanalada de la Cueva Los Grifos, Ocozocoautla, Chiapas*. Cuadernos de Trabajo no. 17. Departamento de Prehistoria, Instituto Nacional de Antropología e Historia, Mexico, D. F.

2001 Cenolítico Superior y Protoneolítico (7000-2500 A. C.). *Arqueología Mexicana* 9(52):52-57.

García Gil, G.
2002 Mapa de Colonización Humana Reciente y Formación del Paisaje Agrario en la Reserva de la Biósfera Calakmul, Campeche, México. Gerardo García Gil, Chetumal.

García Gil, G., J. L. Palacio Prieto and M. A. Ortíz
2002 Reconocimiento Geomorfológico e Hidrográfico de la Reserva de la Biósfera Calakmul, México. *Boletín del Instituto de Geografía, UNAM* (48):7-23.

García Gil, G. and J. M. Pat
2000 Appropriation of Space and Colonization in the Biosphere Reserve in Calakmul, Campeche, Mexico. *Revista Mexicana del Caribe* 10:121-231.

García Heras, M., J. Reyes Trujeque, R. Ruiz Guzmán, M. A. Avilés Escaño, A. Ruiz Conde and P. J. Sánchez Soto
2006 Estudio Arqueométrico de Figurillas Cerámicas Mayas de Calakmul (Campeche, México). *Boletín de la Sociedad Española de Cerámica y Vidrio* 45(4):245-254.

García Moreno, R. and J. Granados
2000 Tumbas Reales de Calakmul. *Arqueología Mexicana* 7(42):28-33.

Gates, G.
1999 Fisiografía, Geología e Hidrología. In *Naturaleza y Cultura en Calakmul, Campeche*, edited by W. J. Folan, C. Sánchez and J. M. García Ortega, pp. 31-39. Centro de Investigaciones Históricas y Sociales, Universidad Autónoma de Campeche, Secretaría de Educación Pública and Proadu, Campeche.

Geovannini, H.
1998 *Edafología y Paisaje de Bonampak, Chiapas, México en el Periodo Clásico.* Licenciatura Thesis, Universidad de las Américas Puebla.

Gibbons, A.
1997 Monte Verde: Blessed, But Not Confirmed. *Science* 5304(28):1256-1257.

Gill, R. B.
2001 Las Grandes Sequías Mayas: Agua, Vida y Muerte. In *Los Investigadores de la Cultura Maya*, pp. 223-238. vol. 9, tomo II. Universidad de Campeche, Campeche.

Gladfelter, B. G.
1977 Geoarchaeology: The Geomorphologist and Archaeology. *American Antiquity* 42:519-538.

Gómez Pompa, A., J. S. Flores and V. Sosa
1987 The "*Pet Kot*": A Man-made Tropical Forest of the Maya. *Interciencia* 12(1):10-15.

González Medrano, F.
2003 *Las Comunidades Vegetales de México.* Secretaría del Medio Ambiente y Recursos Naturales and Instituto Nacional de Ecología, México, D. F.

Graefe, S.
2003 *Crop and Soil Variability in Traditional and Modern Maya Maize Cultivation of Yucatan, Mexico.* Beiheft Nr.75 zu Journal of Agriculture and Rural Development in the Tropics and Subtropics. Kassel University Press GmbH, Kassel.

Grube, N. and S. Martin
2001 La Historia Dinástica de los Mayas. In *Los Mayas. Una Civilización Milenaria*, edited by N.

Grube, pp. 149-171. Könemann Verlagsgesellschaft mbH, Colonia.

Gruhn, R. and A. Bryan
1977 Los Tapiales, a Paleo-Indian Campsite in the Guatemalan Highlands. *Proceedings of the American Philosophical Society* 121(3):235-273.

Gunn, J. D., W. J. Folan and H. R. Robichaux
1995 A Landscape Analysis of the Candelaria Watershed in Mexico: Insights into Paleo-climates Affecting Upland Horticulture in the Southern Yucatan Peninsula Semi-karst. *Geoarchaeology: An International Journal* 10(1):3-42.

Gunn, J. D., J. E. Foss, W. J. Folan, M. d. R. Domínguez Carrasco and B. B. Faust
2002 Bajo Sediments and the Hydraulic System of Calakmul, Campeche, Mexico. *Ancient Mesoamerica* 13(2):297-315.

Gurri, F. D.
2006 25 Años de Colonización. Sobreviviendo y Garantizando el Futuro en Calakmul. *Ecofronteras* 28:2-6.

Gutiérrez, C.
2000 *Listado Florístico Actualizado del Estado de Campeche*. Centro de Investigaciones Históricas y Sociales, Universidad Autónoma de Campeche, Campeche.

Hammond, N.
2001 Los Orígenes de la Cultura Maya y la Formación de Comunidades Rurales. In *Los Mayas. Una Civilización Milenaria*, edited by N. Grube, pp. 35-47. Könemann Verlagsgesellschaft mbH, Colonia.

Hansen, R. D.
2001 Primeras Ciudades. Urbanización Incipiente y Formación de Estados en las Tierras Bajas Mayas. In *Los Mayas. Una Civilización Milenaria*, edited by N. Grube, pp. 50-65. Könemann Verlagsgesellschaft mbH, Colonia.

2004 El Mirador, Guatemala. El Apogeo del Preclásico en el Area Maya. *Arqueología Mexicana* 11(66):28-33.

Hansen, R. D., S. Bozarth, J. S. Jacob, D. Wahl and T. Schreiner
2002 Climatic and Environmental Variability in the Rise of Maya Civilization. *Ancient Mesoamerica* 13(2):273-295.

Harlan, J. R.
1997 Distribution of Agricultural Origins. A Global Perspective. In *The Origins of Agriculture and Crop Domestication. The Harlan Symposium.*, edited by A. B. Damania, J. Valkoun, G. Willcox and C. O. Qualset, html online version http://www.ipgri.cgiar.org/publications/HTMLPublications/47/index.htm. ICARDA, Aleppo, Syria.

Harris, M.
1979 *Cultural Materialism*. Random House, New York.

Harrison, P. D.
1996 Settlement and Land Use in the Pulltrouser Swamp Archaeological Zone, Northern Belize. In *The Managed Mosaic: Ancient Maya Agriculture*, edited by S. L. Fedick, pp 177-190. University of Utah Press, Salt Lake City.

Harrison, P. D. and B. L. I. Turner (editors)
1978 *Prehispanic Maya Agriculture*. University of New Mexico Press, Albuquerque.

Haug, G. H., D. Günther, L. Peterson, D. M. Sigman, K. A. Hughen and B. Aeschlimann
2003 Climate and the Collapse of Maya Civilization. *Science* 299:1731-1735.

Hawkes, J. G.
1998 Back to Vavilov: Why Were Plants Domesticated in Some Areas and Not in Others? In *The Origins of Agriculture and Crop Domestication. The Harlan Symposium*, edited by A. B. Damania, J. Valkoun, G. Willcox and C. O. Qualset, online version http://www.ipgri.cgiar.org/publications/HTMLPublications/47/index.htm. ICARDA, Aleppo, Syria.

Healey, P., J. Lambert, J. T. Armason and R. Hebda
1983 Caracol, Belize: Evidence of Ancient Maya Agricultural Terraces. *Journal of Field Archaeology* 10:397-410.

Heckman, J.
n.d. *Soil Fertility Test Interpretation Fact Sheet FS719*. Rutgers Cooperative Extension. The State University of New Jersey, New Brunswick.

Hellmuth, N.
1977 Cholti-Lacandon (Chiapas) and Peten-Ytza Agriculture, Settlement Pattern and Population. In *Social Process in Maya Prehistory: Studies in Honor of Sir Eric Thompson*, edited by N. Hammond, pp. 421-428. Academic Press, London.

Hernández Xolocotzi, E.
1959 La Agricultura. In *Los Recursos Naturales del Sureste y su Aprovechamiento*, edited by

Enrique. Beltrán, pp. 1-38, vol. 3. IMRNRAC, México D. F.

Hester, J.
1954 *Natural and Cultural Bases of Ancient Maya Subsistence Economy.* PhD Dissertation, University of California Los Angeles.

Hester, T. R., H. B. Iceland, D. B. Hudler and H. J. Shafer
1995 The Colha Preceramic Project: A Status Report. *Texas Papers on Latin America* (95-03):1-10.

Hodder, I. and S. Hutson
2003 *Reading the Past: Current Approaches to Interpretation in Archaeology.* Third ed. Cambridge University Press, Cambridge.

Hodell, D. A., J. H. Curtis and M. Brenner
1995 Possible Role of Climate in the Collapse of Classic Maya Civilization. *Nature* 375(1):391-394.

Hoffecker, J. F., W. R. Powers and T. Goebel
1993 The Colonization of Beringia and the Peopling of the New World. *Science* 259:46-53.

Ingold, T.
1986 *The Appropriation of Nature. Essays in Human Ecology and Social Relations.* Manchester University Press, Manchester.

Instituto de Ciencias de la Universidad Autónoma de Puebla (ICUAP)
n. d. *Manual de Suelos.* Departamento de Investigación en Ciencias Agrícolas, Benemérita Universidad Autónoma de Puebla, Puebla.

Instituto Nacional de Ecología (INE)
2000 *Programa de Manejo de La Reserva de la Biósfera de Calakmul.* Secretaría de Medio Ambiente y Recursos Naturales, México, D.F.

Instituto Nacional de Estadística Geografía e Informática (INEGI)
1985 Carta Uso de Suelo y Vegetación Chetumal E16-4-7. Escala 1:250 000. First ed. Instituto Nacional de Estadística Geografía e Informática, Aguascalientes.

1987 Carta Topográfica Calakmul E16A81. Escala 1:50000. INEGI, México, D. F.

1988 Carta Hidrológica de Aguas Superficiales Chetumal E16-4-7. Escala 1:250 000. First ed. Instituto Nacional de Estadística Geografía e Informática, México, D. F.

1989 Carta Hidrológica de Aguas Subterráneas Chetumal 16-4-7. Escala 1:250 000. First ed. Instituto Nacional de Estadística Geografía e Informática, México, D. F.

1998 Diccionario de Datos Edafológicos. Escala 1:1 000 000 (Vectorial), http://mapserver.inegi.gob.mx/geografia/espanol/normatividad/diccio/eda_1000.pdf.

1999 Carta Geológica Chetumal E16-4-7. Escala 1:250 000. First ed. Instituto Nacional de Estadística Geografía e Informática, Aguascalientes.

2001a Fotografías Aéreas Escala 1:75 000, Zona E16-4-7, Línea 249. INEGI, México, D. F.

2001b Digital Orthophotographs E16A81B and E16A81E. INEGI, México, D. F.

2002 Carta Topográfica Chetumal E16-4-7. Escala 1:250 000. INEGI, Aguascalientes.

International Land Development Consultants (ILACO)
1981 *Agricultural Compendium for Rural Development in the Tropics and Subtropics.* Elsevier Scientific Publishing Company, Amsterdam.

IRENAT and Colegio de Posgraduados
n.d. *Manual de Procedimientos Analíticos para Análisis de Suelos y Plantas del Laboratorio de Fertilidad de Suelos.* Sociedad Mexicana de la Ciencia del Suelo, México, D. F.

Isendahl, C.
2002 *Common Knowledge: Lowland Maya Urban Farming at Xuch.* Doctoral Dissertation, Uppsala University.

Islebe, G. A., H. Hooghiemstra, M. Brenner, J. H. Curtis and D. A. Hodell
1996 A Holocene Vegetation History from Lowland Guatemala. *The Holocene* 6(3):265-271.

Issac Márquez, R.
2004 Explorando la Perspectiva Campesina de la Agroforestería en la Reserva de la Biósfera de Calakmul. *Universidad y Ciencia* 20(40):39-54.

Jacob, J. S.
1995 Archaeological Pedology in the Maya Lowlands. In *Pedological Perspectives in Archaeological Research.*, pp. 51-79. Soil Science Society of America Special Publication 44. Soil Science Society of America, Madison.

Jenny, H.
1941 *Factors of Soil Formation.* McGraw Hill, New York.

Kirchhoff, P.

1943 Mesoamerica. Its Geographic Limits, Ethnic Composition and Cultural Characteristics. *Acta Americana* 2:92-107.

Klingebiel, A. A. and P. H. Montgomery
1961 *Land Capability Classification.* Agriculture Handbook 210. United States Department of Agriculture, Washington, D. C.

Knapp, A. B. and W. Ashmore
1999 Archaeological Landscapes: Constructed, Conceptualized, Ideational. In *Archaeologies of Landscape. Contemporary Perspectives*, edited by A. B. Knapp and W. Ashmore, pp 1-30. Blackwell Publishers, Oxford.

Koontz, R., K. Reese-Taylor and A. Headrick (editors)
2001 *Landscape and Power in Ancient Mesoamerica.* Westview Press, Colorado.

LaMotte Company
n.d. *Soil Texture Unit.* LaMotte Company, Maryland.

Lentz, D. L.
1999 Plant Resources of the Ancient Maya: The Paleoethnobotanical Evidence. In *Reconstructing Ancient Maya Diet*, edited by C. D. White, pp. 3-18. University of Utah Press, Salt Lake City.

2000 *Imperfect Balance: Landscape Transformations in the Precolumbian Americas.* Columbia University Press, New York.

León, R.
2003 *Manual Edafológico de Campo.* Second ed. Textos Universitarios. Universidad de Veracruz, Xalapa.

Leyden, B. W., M. Brenner and B. Dahlin
1998 Cultural and Climatic History of Cobá, a Lowland Maya City in Quintana Roo, México. *Quaternary Research* 49:111-122.

Leyden, B. W., M. Brenner, T. J. Whitmore, J. H. Curtis, D. R. Piperno and B. Dahlin
1996 A Record of Long- and Short-Term Climatic Variation from Northwest Yucatan: Cenote San Jose Chulchaca. In *The Managed Mosaic: Ancient Maya Agriculture*, edited by S. L. Fedick, pp 30-50. University of Utah Press, Salt Lake City.

Liendo, R.
2002 *La Organización de la Producción Agrícola en un Centro Maya del Clásico: Patrón de Asentamiento en la Región de Palenque, Chiapas.* Translated by Concepción Obregón. Instituto Nacional de Antropología e Historia, México, D. F.

Linteau, J. P.
1996 *Agroforestry for Sustainable Development in Calakmul.* Field report on work in the Calakmul Model Forest (Oct.-Dec. 1995) prepared for the International Development Research Center (IDRC).

Lobato, R.
1988 Terrazas Prehispánicas en la Región del Río Usumacinta y su Importancia en la Agricultura Maya. *Estudios de Cultura Maya* (17):19-58.

Long, A. B., B. F. Benz, D. J. Donahue, A. Jull and L. J. Toolin
1989 First Direct AMS Dates on Early Maize From Tehuacan Mexico. *Radiocarbon* 31:1035–1040.

López Austin, A. and L. López-Luján
2000 La Periodización de la Historia Mesoamericana. *Arqueología Mexicana* 8(43):14-21.

López Luján, L.
1993 La Cuenca de México durante la Época Mexica. In *Atlas Histórico de Mesoamérica*, edited by L. Manzanilla and L. López-Luján, pp. 148-152. Larousse Ediciones, México, D.F.

Lorenzo, J. and L. Mirambell (editors)
1986 *Tlapacoya: 35 000 años de Historia del Lago de Chalco.* Instituto Nacional de Antropología e Historia, México, D.F.

Lugo, J. I.
1988 *Elementos de Geomorfología Aplicada.* Instituto de Geografía. Universidad Nacional Autónoma de México, México, D. F.

Lundell, C. L.
1933 Archaeological Discoveries in the Maya Area. *Proceedings of the American Philosophical Society* 72(3):147-179.

1934 *Preliminary Sketch of the Phytogeography of the Yucatan Peninsula.* Publication 436, pp. 257-321. Carnegie Institute of Washington, Washington, D. C.

1937 *The Vegetation of Petén.* Carnegie Institution of Washington, Washington, D. C.

MacNeish, R. S.
1964 *El Origen de la Agricultura Visto desde Tehuacán.* Publicación 16. Departamento de Prehistoria, Instituto Nacional de Antropología e Historia, México, D. F.

Mallarino, A. and J. E. Sawyer
1999 *Interpreting Mehlich-3 Soil Test Results. Electronic Version.* Iowa State University.

Manzanilla, L.
1993a El Mundo Clásico Mesoamericano. In *Atlas Histórico de Mesoamérica*, edited by L. Manzanilla and L. López-Luján, pp. 74-76. Larousse Ediciones, México, D. F.

1993b La Ciudad de Teotihuacan. In *Atlas Histórico de Mesoamérica*, edited by L. Manzanilla and L. López-Luján, pp 81-84. Larousse Ediciones, México, D. F.

Marcus, J.
1973 Territorial Organization of the Lowland Classic Maya. *Science* 180:911-916.

2001 Clásico Tardío (600-900 d. C.). *Arqueología Mexicana* 8(48):20-29.

Martin, S.
1996 Calakmul y el Enigma del Glifo Cabeza de Serpiente. *Arqueología Mexicana* 3(18):42-45.

1997 The Painted King List: A Commentary on Codex-Style Dynastic Vases. In *The Maya Vase Book*, pp. 847-967. vol. 5. Kerr Associates.

2000 Los Señores de Calakmul. *Arqueología Mexicana* 7(42):40-45.

2005 De Serpientes y Murciélagos: Identidades Cambiantes en Calakmul. *Pari Journal* 6(2):5-1.

Martin, S. and N. Grube
1995 Maya Superstates. *Archaeology* 48(6):41-46.

2002 Crónica de los Reyes y Reinas Mayas: La Primera Historia de las Dinastías Mayas. Planeta, México, D. F.

Martínez, E.
1999 *La Vegetación en Calakmul*. Informe Técnico. Pronatura Península de Yucatán, Mérida.

Martínez, E. and C. Galindo-Leal
2002 La Vegetación de Calakmul, Campeche, México: Clasificación, Descripción y Distribución. *Bol. Soc. Bot. México* 71:7-32.

Martínez, E., C. H. Ramos and M. Sosa
1999 Lista Florística Preliminar de Calakmul de Arboles y Plantas Melíferas. Archivo Digital. Herbario Nacional. Universidad Nacional Autónoma de México.

Martínez, E., M. Sousa and C. H. Ramos
2001 Listados Florísticos de México. XXII. Región de Calakmul Campeche. Instituto de Biología, Universidad Nacional Autónoma de México, México, D. F.

Marx, E. S., J. Hart and R. G. Stevens
1999 *Soil Test Interpretation*. Oregon State University Extension Service, Corvallis.

Masson, M. A., C. Peraza Lope and T. S. Hare
2002 Surface Evidence of Mayapán's Political Economy. Paper presented at the Society for American Archaeology Meetings, Denver.

Matheny, R. T.
1976 Maya Lowland Hydraulic Systems. *Science* 193:639-646.

1986 Investigaciones en El Mirador, Petén, Guatemala. *National Geographic Research* (2):332-353.

May Hau, J., R. Cohuoh Muñoz, R. González Heredia and W. J. Folan
1990 *El Mapa de Calakmul, Campeche*. Centro de Investigaciones Históricas y Sociales, Universidad Autónoma de Campeche, Campeche.

McClung de Tapia, E.
1993 La Domesticación de las Plantas Alimenticias. El Origen de la Agricultura. In *Atlas Histórico de Mesoamérica*, edited by L. Manzanilla and L. López-Luján, pp. 45-49. Ediciones Larousse, México, D. F.

1997 La Domesticación del Maíz. *Arqueología Mexicana* 5(25):34-39.

McKillop, H.
1994 Ancient Maya Tree Cropping. *Ancient Mesoamerica* 5:129-140.

Miller, N. F. and K. L. Gleason (editors)
1994 *The Archaeology of Garden and Field*. University of Pennsylvania Press, Philadelphia.

Mirambell, L.
2001 Arqueolítico y Cenolítico Inferior (30000 - 7000 A.C.). *Arqueología Mexicana*: 9(52): 46-49.

Miranda, F.
1958 Estudios Acerca de la Vegetación. In *Los Recursos Naturales del Sureste y su Aprovechamiento*, edited by E. Beltrán, pp. 215-271. vol. 2. Instituto Mexicano de Recursos Naturales Renovables, A. C., México, D. F.

Morales, H. and I. Perfecto
2000 Traditional Knowledge and Pest Management in the Guatemalan Highlands. *Agriculture and Human Values* 17(1):49-63.

Morales, J.
1999 Suelos. In *Naturaleza y Cultura en Calakmul, Campeche*, edited by W. J. Folan, C. Sánchez and J. M. García Ortega, pp. 41-49. Centro de

Investigaciones Históricas y Sociales, Universidad Autónoma de Campeche, Secretaría de Educación Pública and Proadu, Campeche.

Morales, J. and S. Magaña
2001 *Fuentes de Impacto, Necesidades de Investigación Científica y Monitoreo en Calakmul, Campeche.* Pronatura Península de Yucatán A. C. and The Nature Conservacy, Mérida.

Morales López, A.
1985 Primer Informe de Salvamento Arqueológico en la Carretera Conhuas - Calakmul, Septiembre - Diciembre 1984. Manuscript in Archivo Técnico, Instituto Nacional de Antropología e Historia, México, D. F.

1987 Arqueología de Salvamento en la Nueva Carretera a Calakmul, Municipio de Champotón, Campeche. *Información* (12):75-109.

Morán, E. F.
1993 *La Ecología Humana de los Pueblos de la Amazonia.* Fondo de Cultura Económica, México, D. F.

Morley, S. G.
1933 The Calakmul Expedition. In *Supplementary Publications*, pp. 30-42. vol. 6. Carnegie Institution of Washington, Washington D. C.

1947 *La Civilización Maya.* Translated by A. Recinos. Fondo de Cultura Económica, México, D. F.

Morrison, B. A. and R. Cózatl Manzano
2003 Initial Evidence for Use of Periphyton as an Agricultural Fertilizer by the Ancient Maya Associated with El Edén Wetland, Northern Quintana Roo, Mexico. In *The Lowland Maya Area. Three Millennia at the Human-Wildland Interface*, edited by A. Gómez-Pompa, M. F. Allen, S. L. Fedick and J. J. Jiménez-Osornio, pp. 401-413. Food Product Press, New York.

Muhs, D. R., R. R. Kautz and J. J. MacKinnon
1985 Soils and the Location of Cacao Orchards at a Maya Site in Western Belize. *Journal of Archaeological Science* 12(2):121-137.

Muir, R.
2000 *The New Reading of the Landscape: Fieldwork in Landscape History.* University of Exeter Press, Exeter.

Munsell Color Company Inc.
1954 *Munsell Soil Color Charts.* Munsell Color Company Inc, Baltimore.

Murtha, T. M. Jr.
2002 *Land and Labor: Classic Maya Terraced Agriculture at Caracol, Belize.* PhD Dissertation, Pennsylvania State University.

Neff, H., B. Arroyo, J. G. Jones, D. M. Pearsall and D. E. Freidel
2001 Nueva Evidencia Pertinente a la Ocupación Temprana del Sur de Mesoamérica. Paper presented at the X Encuentro Internacional Los Investigadores de la Cultura Maya, Campeche.

Nelson, Z. N.
2005 *Settlement and Population at Piedras Negras, Guatemala.* PhD Dissertation, Pennsylvania State University.

Norton, W.
1989 *Explorations in the Understanding of Landscape. A Cultural Geography.* Greenwood Press, New York.

Nuñez, M. G.
1977 *Archaeology Through Soil Chemical Analysis: An Evaluation.* Moniste No. 14. Helsingia Yliopsiton Arkeoligian Laitos, Helsinki.

Orellana, R., G. A. Islebe and C. Espadas
2003 Presente, Pasado y Futuro de los Climas de la Península de Yucatán. In *Naturaleza y Sociedad en el Area Maya. Pasado, Presente y Futuro*, edited by P. Colunga-García Marín and A. Larqué, pp. 37-68. Academia Mexicana de Ciencias and Centro de Investigación Científica de Yucatán, Mérida.

Palacios, E. J.
1933 Informe de Visita al Sitio de Calakmul, Manuscript in the archive of Centro Regional Yucatán, Instituto Nacional de Antropología e Historia, Mérida.

Pasquinucci, M. and F. Trément (editors)
2000 *Non-destructive Techniques Applied to Landscape Archaeology.* Oxbow, Oxford.

Pincemin, S., J. Marcus, L. Florey Folan, W. J. Folan, M. d. R. Domínguez Carrasco and A. Morales López
1998 Extending the Calakmul Dynasty Back in Time: A New Stela from a Maya Capital in Campeche, Mexico. *Latin American Antiquity* 9(4):310-327.

Piperno, D. R. and K. V. Flannery
2001 The Earliest Archaeological Maize (*Zea mays L.*) From Highland Mexico: New Accelerator Mass Spectrometry Dates and their Implications. *Proceedings of the National Academy of Sciences* 98(4):2101-2103.

Piperno, D. R. and D. M. Pearsall

1998 *Origins of Agriculture in the Lowland Neotropics*. Academic Press, San Diego.

Piperno, D. R., A. J. Ranere, I. Holst and P. Hansell
2000 Starch Grains Reveal Early Root Crop Horticulture in the Panamanian Tropical Forest. *Nature* 407:894-897.

Pohl, M.
1990 The Río Hondo Project in Northern Belize. In *Ancient Maya Wetland Agriculture. Excavations at Albion Island*, edited by M. Pohl, pp. 1-19. Westview, Boulder.

Pohl, M. D., K. O. Pope, J. G. Jones, J. S. Jacob, D. R. Piperno, S. D. deFrance, D. L. Lentz, J. A. Gifford, M. E. Danforth and J. K. Josserand
1996 Early Agriculture in the Maya Lowlands. *Latin American Antiquity* 7(4):355-372.

Pool, L.
2001 Dinámica de la Milpa en Yucatán: Una Experiencia de Investigación Participativa. *Ecofronteras* 13:26-28.

Pope, K. O. and B. Dahlin
1989 Ancient Maya Wetland Agriculture: New Insights form Ecological and Remote Sensing Research. *Journal of Field Archaeology* 16(1):87-106.

Pope, K. O., M. D. Pohl, J. G. Jones, D. L. Lentz, C. von Nagy, F. J. Vega and I. R. Quitmyer
2001 Origin and Environmental Setting of Ancient Agriculture in the Lowlands of Mesoamerica. *Science* 292:1370-1373.

Porta, J., M. López Acevedo and C. Roquero
1994 *Edafología para la Agricultura y el Medio Ambiente*. Ediciones Mundi Prensa, Madrid.

Prado Roque, S. A.
n. d. *Estrategia Preliminar para la Aplicación de la Política de Gestión del Agua por Cuenca en la Región XII, Península de Yucatán*. Gerencia Regional de la Península de Yucatán, Comisión Nacional del Agua, Mérida.

Puleston, D. E.
1978 Terracing, Raised Fields, and Tree Cropping in the Maya Lowlands: A New Perspective on the Geography of Power. In *Pre-Hispanic Maya Agriculture*, edited by P. D. Harrison and B. L. I. Turner, pp. 225-245. University of New Mexico Press, Albuquerque.

1982 The Role of Ramon in Maya Subsistence. In *Maya Subsistence : Studies in Memory of Dennis E. Puleston.*, edited by K. V. Flannery, pp. 349-366. Studies in Archaeology. Academic Press, New York.

Pyburn, K. A.
1996 Political Economy of Ancient Maya Land Use: The Road to Ruin. In *The Managed Mosaic: Ancient Maya Agriculture and Resource Use*, edited by S. L. Fedick, pp. 236-247. University of Utah Press, Salt Lake City.

2003 The Hydrology of Chau Hiix. *Ancient Mesoamerica* 14:123-129.

Recinos, A., D. Goetz and S. G. Morley
1954 Popol Vuh. Translated by D. Goetz and S. G. Morley, from Adrián Recino's translation from Quiché into Spanish,. http://www.sacred-texts.com/nam/maya/pvgm/, accessed September 7th 2005.

Reents Budet, D.
1998 The Ceramic Chemistry of Maya Kingship: New Research at Calakmul, Mexico. Paper presented at the VI Annual Maya Week of the Archaeology Institute, University of California, Los Angeles.

Rehm, G.
1994 *Soil Cation Ratios for Crop Production*. North Central Regional Extension Publication 533. University of Minnesota, St. Paul.

Research Branch
1976 *Glossary of Terms in Soil Science*. Publication 1459. Canada Department of Agriculture, Ottawa.

Rice, D. S.
1996 Paleoliminological Analysis in the Central Petén, Guatemala. In *The Managed Mosaic: Ancient Maya Agriculture and Resource Use*, edited by S. L. Fedick, pp. 193–206. University of Utah Press, Salt Lake City.

Rodríguez, O.
2000 La Gran Plaza de Calakmul. *Arqueología Mexicana* 7(42):22-27.

2003 *La Arquitectura Petén en Calakmul: Una Comparación Regional.* Report submitted to Foundation for the Advancement of Mesoamerican Studies. http://www.famsi.org/reports/02070es/index.html

Rosenmeier, M. F., D. A. Hodell, M. Brenner and J. H. Curtis
2002 A 4000-Year Lacustrine Record of Environmental Change in the Southern Maya Lowlands, Peten, Guatemala. *Quaternary Research* 57:183-190.

Rossignol, J. and L. Wandsnider (editors)
1992 *Space, Time, and Archaeological Landscapes.* Plenum Press, New York.

Roys, R.
1943 *The Indian Background of Colonial Yucatan.* vol. 548. The Carnegie Institution of Washington, Washington, D. C.

Ruppert, K. and J. H. Denison Jr.
1943 *Archaeological Reconnaissance in Campeche, Quintana Roo and Peten.* Publication 543. The Carnegie Institution of Washington, Washington, D. C.

Sage, R. F.
1995 Was Low Atmospheric CO_2 During the Pleistocene a Limiting Factor for the Origin of Agriculture? *Global Change Biology* 1:93-106.

Sánchez, P. A. and T. J. Logan
1992 Myths and Science About the Chemistry and Fertility of Soils in the Tropics. In *Myths and Science of Soils of the Tropics*, edited by R. Lal and P. A. Sánchez, pp. 35 – 47. SSSA Special Publication 29. Soil Science Society of America Inc, Madison.

Santley, R.
1990 Demographic Archaeology in the Maya Lowlands. In *Precolumbian Population History in the Maya Lowlands*, edited by T. P. Culbert and D. S. Rice, pp. 325-343. University of New Mexico Press, Albuquerque.

Sauer, C.
1925 The Morphology of Landscape. *University of California Publications in Geography* 2(2):19-53.

1940 *Hacia una Geografía Histórica. Discurso a la Asociación Norteamericana de Geógrafos.* www.colorado.edu/geography, Baton Rouge.

Savage, S. H.
1990 GIS in Archaeological Research. In *Interpreting Space: GIS and Archaeology*, edited by K. M. S. Allen, S. W. Green and E. B. W. Zubrow, pp. 22-33. Taylor and Francis, London.

Scarborough, V., M. Becher, J. L. Baker, G. Harris and F. Valdez Jr.
1995 Water and Land at the Ancient Maya Community of La Milpa. *Latin American Antiquity* 6(2):98-119.

Scarborough, V. L., R. P. Connolly and S. P. Ross
1994 The Pre-Hispanic Maya Reservoir System at Kinal, Peten, Guatemala. *Ancient Mesoamerica* 5:97-106.

Scarborough, V. L. and G. G. Gallopin
1991 A Water Storage Adaptation in the Maya Lowlands. *Science* 251:658-662.

Schefller, T. E.
2001 *El Gigante Rock Shelter: Archaic Mesoamerica and Transitions to Settled Life.* Report submitted to FAMSI. http://www.famsi.org/reports/00071/index.html.

Schlesinger, V.
2004 *Animals and Plants of the Ancient Maya.* University of Texas Press, Austin.

Schuster, A. M. H.
1997 The Search of Site Q. *Archaeology* 50(5):42.

Secretaría de Programación y Presupuesto (SPP)
1980 Carta de Climas 1:000 000. Secretaría de Programación y Presupuesto, Mérida.

1981 Carta Edafológica 1:1 000 000. Secretaría de Programación y Presupuesto, México, D. F.

Secretaría del Medio Ambiente Recursos Naturales y Pesca (SEMARNAP)
1995 *Reservas de la Biósfera y Otras Areas Protegidas de México.* Secretaría del Medio Ambiente Recursos Naturales y Pesca, México, D. F.

Serra, M. C.
1993 El Preclásico. La Etapa Aldeana. In *Atlas Histórica de Mesoamérica,* edited by L. Manzanilla and L. López-Luján. Pp.50-56. Larousse Ediciones, México, D. F.

Sever, T. L. and D. E. Irwin
2003 Remote-Sensing Investigation of the Ancient Maya in the Peten Rainforest of Northern Guatemala. *Ancient Mesoamerica* 14:113-122.

Shackley, M.
1981 *Environmental Archaeology.* George and Allen Unwin, London.

Sharpley, A. N., T. C. Daniel and D. R. Edwards
1994 Phosphorus Movement in the Landscape. *Journal Production Agriculture* 6(4):492-500.

Siemens, A. H. and D. E. Puleston
1972 Ridged Fields and Associated Features in Southern Campeche: New Perspectives on the Lowland Maya. *American Antiquity* 37:228-239.

1982 Prehispanic Cultural Use of the Wetlands of Northern Belize. In *Maya Subsistence Studies in Memory of Dennis E. Puleston*, edited by K. V. Flannery, pp. 205–225. Academic Press, New York.

Sistemas de Información Geográfica (SIGSA)

2004 Información Cartográfica Digital. Estado de Campeche E16U181, E16U182, Escala 1:20 000. SIGSA, México, D. F.

Sivakumar, M. V. K., A. Manu, V. S.M and E. T. Kanemasu
1992 Relation between Climate and Soil Productivity in the Tropics. In *Myths and Science of Soils of the Tropics*, edited by R. Lal and P. A. Sánchez, pp. 91-119. SSSA Special Publication No. 29. SSSA and ASA, Madison.

Smith, B. D.
1997 The Initial Domestication of *Cucurbita pepo* in the Americas 10,000 Years Ago. *Science* 276:932-934.

Sombroek, W. G. and R. F. van de Weg
1980 *Annual Report*. International Soil Museum, Wageningen.

Stahl, P. W.
1996 Holocene Biodiversity: An Archaeological Perspective from the Americas. *Annual Review of Anthropology* 25:105-126.

Steggerda, M.
1941 *Maya Indians of Yucatan*. Publication 531. Carnegie Institution of Washington, Washington, D.C.

Storie, E. R.
1970 *Manual de Evaluación de Suelos*. Centro Regional de Ayuda Técnica, México, D.F.

Stromsvik, G.
1937 Notes on the Metates from Calakmul, Campeche, and from the Mercado, Chichen Itza, Yucatan. In *Contributions to American Anthropology and History*, pp. 121-127. Publication No. 456. vol. 3 (16). Carnegie Institution of Washington, Washington, D. C.

Stuart, D.
2000 Las Nuevas Inscripciones del Templo XIX, Palenque. *Arqueología Mexicana* 8(45):28-33.

2004 The Paw Stone: The Place Name of Piedras Negras, Guatemala. *PARI Journal* 4(3):1-6.

Stuart, D. and S. D. Houston
1994 *Classic Maya Place Names*. Studies in Pre-Columbian Art and Archaeology 33. Dumbarton Oaks, Washington, D. C.

Sugiura, Y.
1993 La Caída del Clásico y el Epiclásico. In *Atlas Histórica de Mesoamérica*, edited by L. Manzanilla and L. López-Luján, pp. 113-117. Larousse Ediciones, México, D. F.

Suyuc, E., B. Balcárcel, F. López and S. Alvarado
n.d. *Excavaciones en el Sitio La Muerta, Cuenca Mirador, Petén*. FAMSI Report No. 03101.

Tamhane, R. V., D. P. Motiramani and Y. P. Bali
1979 *Suelos: Su Quimica y Fertilidad en Zonas Tropicales*. Editorial Diana, Mexico, D. F.

Tejeda, S., V. Tiesler, W. J. Folan and M. Coyoc Ramírez
2001 Nutrición y Estilo de Vida en Calakmul. In *Los Investigadores de la Cultura Maya*, pp. 336-340. vol. 9, tomo II. Universidad Autónoma de Campeche, Campeche.

Terán, C. S. and C. H. Rasmussen
1995 Genetic Diversity and Agricultural Strategy in 16th Century and Present-day Yucatecan Milpa Agriculture. *Biodiversity and Conservation* 4:363-381.

Tilley, C.
1994 *A Phenomenology of Landscape*. Berg, Oxford.

Tozzer, A. M. (editor)
1941 *Landa's Relación de las Cosas de Yucatán. A Translation. Papers of the Peabody Museum of American Archaeology and Ethnology*, 18. Harvard University, Cambridge.

Trigger, B.
1992 *Historia del Pensamiento Arqueológico*. Editorial Crítica, Barcelona.

Turner, B. L. I I.
1983 Constructional Inputs for Major Agrosystems of the Ancient Maya. In *Drained Field Agriculture in Central and South America*, edited by J. P. Darch, pp. 11-26. BAR International Series 189. British Archaeological Reports, Oxford.

1990 Population Reconstruction of the Central Maya Lowlands: 1000 B.C. to A.D. 1500. In *Pre-Columbian Population History in the Maya Lowlands*, edited by T. P. Culbert and D. S. Rice, pp. 63-115. University of New Mexico Press, Albuquerque.

Turner, B. L. I I. and P. D. Harrison
1981 Prehistoric Raised-Field Agriculture in the Maya Lowlands. *Science* 213:399-405.

Turner, B. L. I I. and P. D. Harrison (editors)
1983 *Pulltrouser Swamp: Ancient Maya Habitat, Agriculture, and Settlement in Northern Belize*. University of Texas Press, Austin.

Turner, B. L. I., P. Klepeis and L. C. Schneider
2003 Three Millennia in the Southern Yucatán Peninsula: Implications for Occupancy, Use, and Carrying Capacity. In *The Lowland Maya Area. Three Millennia at the Human-Wildland*

Interface, edited by A. Gómez-Pompa, K. M. S. Allen, S. L. Fedick and J. J. Jiménez-Osornio, pp. 361-387. Food Products Press, New York.

Ucán, E., L. Ortega, J. Ortiz, J. Tun and J. S. Flores
1999 Listado Florístico de la Reserva de la Biósfera Calakmul. In *Naturaleza y Cultura en Calakmul, Campeche*, edited by W. J. Folan, J. M. García Ortega and M. Sánchez, pp. 139-159. Centro de Investigaciones Históricas y Sociales, Universidad Autónoma de Campeche, Campeche.

United Nations
1997 *Glossary of Environment Statistics*. Studies in Methods, Series F, No. 67. United Nations, New York.

United States Department of Agriculture
1962 *Soil Survey Manual*. U. S. Department of Agriculture Handbook 18. U. S. Government Printing Office, Washington, D. C.

United States Department of Agriculture Soil Survey Division Staff
1993 *Soil Survey Manual*. Third ed. USDA Handbook 18. United States Department of Agriculture, Washington, D.C.

Universidad de Quintana Roo
2000 *Programa Estatal de Ordenamiento Territorial, Quintana Roo*. Universidad de Quintana Roo, Chetumal.

Valadez, R.
1999 Los Animales Domésticos. *Arqueología Mexicana* 6(35):32-39.

Vázquez López, V. A.
2006 Pintura Mural y Arquitectura como Medios de Transmisión Ideológica: La Acrópolis Chiik Naab′. In *Memorias del XV Encuentro Internacional Los Investigadores de la Cultura Maya*, en prensa, vol. 14. Universidad Autónoma de Campeche, Campeche.

Velásquez, E.
2004 Los Escalones Jeroglíficos de Dzibanche. In *Los Cautivos de Dzibanche*, edited by E. Nalda, pp. 79-105. Instituto Nacional de Antropología e Historia, México, D.F.

Vink, A. P. A.
1968 *Aerial Surveys and Integrated Studies*. UNESCO, Paris.

Vita Finzi, D. and E. S. Higgs
1970 Prehistoric Economy in the Mount Carmel Area of Palestine: Site Catchment Analysis. *Proceedings of the Prehistoric Society* 36:1-37.

Wagner, H. O.
1968 Subsistence Potential and Population Density of the Maya on the Yucatan Peninsula and Causes for the Decline on Population in the Fifteenth Century. In *Verhandlungen des XXXVIII Internnational Amerikanisten Kongresses*, pp. 179-196. Komissionverlag, Klaus Renner, München.

Wagstaff, J. M. (editor)
1987 *Landscape and Culture. Geographical and Archaeological Perspectives*. Basil Blackwell, Oxford.

Wahl, D.
2005 Environmental Change and Prehistoric Agriculture in the Mirador Basin. Report submitted to FAMSI, http://www.famsi.org/reports/01071/index.html.

Wahl, D., R. Byrnea, T. Schreinera and R. Hansenb
2006 Holocene Vegetation Change in the Northern Peten and its Implications for Maya Prehistory. *Quaternary Research* 65(3):380-389.

Webster, D.
2002 *The Fall of the Ancient Maya*. Thames and Hudson, London.

Whitmore, T. J., M. Brenner, J. H. Curtis, B. Dahlin and B. W. Leyden
1996 Holocene Climate and Human Influences on Lakes of the Yucatan Peninsula, Mexico. *The Holocene* (6):273–287.

Williams, G.
1966 *Adaptation and Natural Selection*. Princeton University Press, Princeton.

Wingard, J. D.
1992 *The Role of Soils in the Development and Collapse of Classic Maya Civilization at Copán, Honduras*, Pennsylvania State University.

1996 Interactions between Demographic Processes and Soil Resources in the Copan Valley, Honduras. In *The Managed Mosaic: Ancient Maya Agriculture and Resource Use*, edited by S. L. Fedick, pp. 207-235. University of Utah Press, Salt Lake City.

Wiseman, F.
1978 Agricultural and Historical Ecology of the Maya Lowlands. In *Prehispanic Maya Agriculture*, edited by P. D. Harrison and B. L. I. Turner, pp. 65-115. University of New Mexico, Albuquerque.

Zapata, A.

REFERENCES

1985 Los Chultunes de Calakmul. Informe de Campo. Universidad Autónoma del Sudeste and Instituto Nacional de Antropología e Historia, Campeche.

Zonneveld, I. S.
1978 *Land Evaluation and Landscape Science.* ITC Textbook VII-4. Enschede, Netherlands.

www.ingramcontent.com/pod-product-compliance
Lightning Source LLC
LaVergne TN
LVHW071217210926
848482LV00015B/834